-Presents-

REVIEW MANUAL

WITH

300 TOTAL REVIEW QUESTIONS

180 TOPICALLY ORGANIZED QUESTIONS

AND

120 QUESTIONS CONTAINED IN

TWO PRACTICE EXAMS

Published by: JD's Regents Preparation, LLC.

Cover illustration by James A. Stiehl

Printed in the United States of America
ISBN: 978-1-939246-70-7

Table of Contents

ALGEBRA

Pre-Algebra 1
25 Questions

Algebra 1 10
18 Questions

Algebra 2 17
14 Questions

Algebra 3 24
14 Questions

Algebra 4 30
7 Questions

GEOMETRY

Geometry 1 33
22 Questions

Geometry 2 45
11 Questions

Geometry 3 50
23 Questions

Geometry 4 60
15 Questions

FUNCTIONS

Function 1 67
6 Questions

Function 2 70
9 Questions

PROBABILITY

Probability 76
7 Questions

STATISTICS

Statistics 80
10 Questions

ACT PRACTICE EXAM 1 84
60 Questions

ACT PRACTICE EXAM 2 112
60 Questions

SOLUTIONS 141

PREALGEBRA

1. 12% of 90 is $\frac{1}{5}$ of what number?

PA 04

   a. 2.16
   b. 27
   c. 54
   d. 540
   e. 5,400

2. If 60% of a given number is 15, then what is 35% of the given number?

PA 01

   a. 5.25
   b. 7.25
   c. 8.75
   d. 9
   e. 21

3. Kyle's current hourly wage for working at Many Miles Garage is \$14.00. Kyle was told that at the beginning of next month, his new hourly wage will be an increase of 8% of his current hourly wage. What will be Kyle's new hourly wage?

PA 03

   a. \$14.08
   b. \$14.80
   c. \$15.12
   d. \$22.00
   e. \$24.20

4. To get a scuba diving license, an applicant must pass a written test and an under-water diving test. Past records show that 70% of the applicants pass the written test and 90% of those who pass the written test pass the under-water diving test. Based on these figures, how many applicants in a random group of 200 applicants would you expect to get their scuba diving license?

PA 24

a. 20
b. 108
c. 126
d. 140
e. 180

5. The product of 4 and 200% of 2 has the same value as which of the following calculations?

PA 02

a. 100% of 8
b. 150% of 8
c. 300% of 8
d. 300% of 4
e. 400% of 4

6. $|9(-4) + 5(2)| = ?$

PA 05

a. −46
b. −26
c. 26
d. 29
e. 46

7. $-5\,|-3 + 9| = ?$

PA 09

a. -60
b. -30
c. 1
d. 30
e. 60

8. $|8 - 5| - |5 - 8| = ?$

PA 07

a. -6
b. -4
c. -2
d. 0
e. 6

9. In which of the following are $\frac{2}{3}, \frac{3}{5}, and\ \frac{5}{9}$ arranged in ascending order?

PA 06

a. $\frac{5}{9} < \frac{3}{5} < \frac{2}{3}$

b. $\frac{5}{9} < \frac{2}{3} < \frac{3}{5}$

c. $\frac{2}{3} < \frac{3}{5} < \frac{5}{9}$

d. $\frac{3}{5} < \frac{2}{3} < \frac{5}{9}$

e. $\frac{2}{3} < \frac{5}{9} < \frac{3}{5}$

10. What is the least common multiple of 40, 70, and 90 ?

PA 08

a. 70
b. 200
c. 252
d. 2,520
e. 25,200

11. What fraction lies exactly halfway between $\frac{3}{4}$ *and* $\frac{4}{5}$?

PA 010

a. $\frac{2}{3}$

b. $\frac{6}{7}$

c. $\frac{3}{20}$

d. $\frac{17}{25}$

e. $\frac{31}{40}$

12. Tony's favorite quiche recipe requires 5 eggs and makes 8 servings. Tony will modify the recipe by using 10 eggs and increasing all other ingredients in the recipe proportionally. What is the total number of servings the modified recipe will make?

PA 011

a. 8
b. 10
c. 12
d. 16
e. 18

PA 23

13. In teaching a lesson on the concept of fourths, Ms. Chow uses a divide-and-set-aside procedure. She starts with a certain number of poker chips, divides them into 4 equal groups, and sets 1 group aside to illustrate $\frac{1}{4}$. She repeats the procedure by taking the chips she had NOT set aside, dividing them into 4 equal groups, and setting 1 of these groups aside. If Ms. Chow wants to be able to complete the divide-and-set-aside procedure at least 3 times (without breaking and of the chips into pieces), which of the following is the minimum number of poker chips she can start with?

    a. 16
    b. 24
    c. 48
    d. 64
    e. 256

PA 012

14. Anna walked at a rate of 5 miles per hour for 15 minutes and then walked at a rate of 3 miles per hour for 5 minutes. Which of the following gives the average rate, in miles per hour, at which she walked over this 20-minute period?

    a. $\frac{3}{4}$
    b. $\frac{4}{3}$
    c. $\frac{7}{2}$
    d. 4
    e. $\frac{9}{2}$

15. What is the sum of the first 5 terms of the arithmetic sequence in which the 6th term is 10 and the 10th term is 16?

PA 13

a. 16.5
b. 19
c. 24
d. 27.5
e. 37.5

16. Which of the following expressions is equivalent to $(x+4)^{-10}$?

PA 14

a. $-x^{10} - 4^{10}$

b. $-10x - 40$

c. $\frac{1}{x^{10}} + \frac{1}{4^{10}}$

d. $\frac{1}{(4x)^{10}}$

e. $\frac{1}{(x+4)^{10}}$

17. Which of the following matrices is equal

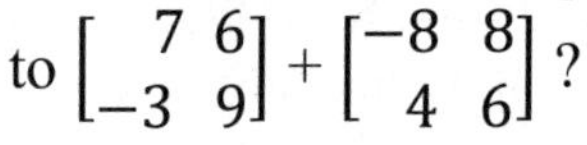

to $\begin{bmatrix} 7 & 6 \\ -3 & 9 \end{bmatrix} + \begin{bmatrix} -8 & 8 \\ 4 & 6 \end{bmatrix}$?

PA 17

a. $\begin{bmatrix} -1 & 14 \\ 1 & 15 \end{bmatrix}$

b. $\begin{bmatrix} -1 & 14 \\ 7 & 15 \end{bmatrix}$

c. $\begin{bmatrix} 15 & 14 \\ 7 & 15 \end{bmatrix}$

d. $\begin{bmatrix} 13 & 0 \\ 6 & 10 \end{bmatrix}$

e. $\begin{bmatrix} -13 & 68 \\ 47 & 6 \end{bmatrix}$

18. Simplify $\frac{8.4\ x\ 10^{-5}}{1.2\ x\ 10^{-9}}$

PA 16

a. 7.0×10^{4}
b. 7.0×10^{-4}
c. 7.0×10^{-1}
d. 7.2×10^{14}
e. 7.2×10^{4}

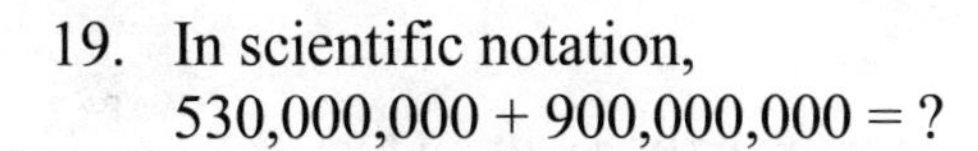
19. In scientific notation,
530,000,000 + 900,000,000 = ?

PA 15

a. 1.43×10^{-9}
b. 1.43×10^{7}
c. 1.43×10^{8}
d. 1.43×10^{9}
e. 143×10^{15}

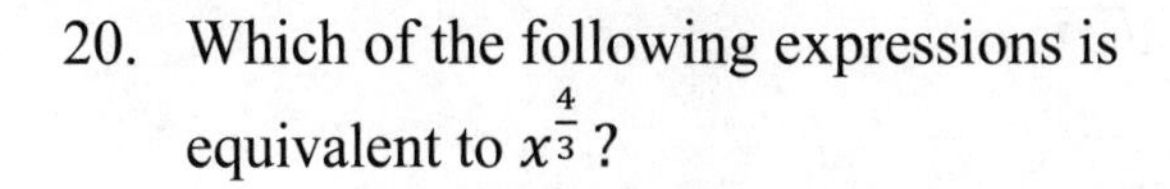
20. Which of the following expressions is equivalent to $x^{\frac{4}{3}}$?

PA 18

a. $\frac{x^4}{3}$

b. $\frac{x\ (4)}{3}$

c. $\sqrt[4]{x^3}$

d. $\sqrt[3]{x^4}$

e. $\sqrt[3]{x}$

21. What is the smallest integer greater than $\sqrt{72}$?

PA 19

a. 3
b. 8
c. 9
d. 11
e. 37

22. Which of the following expression, when evaluated equals an irrational number?

PA 21

a. $\frac{\sqrt{2}}{\sqrt{32}}$

b. $\frac{\sqrt{32}}{\sqrt{2}}$

c. $\left(\sqrt{32}\right)^2$

d. $\sqrt{32} \times \sqrt{2}$

e. $\sqrt{32} + \sqrt{2}$

23. The number 1045 is the product of the prime numbers 5, 11, and 19. Knowing this, what is the prime factorization of 20,900?

PA 20

a. $2 \cdot 5 \cdot 10 \cdot 19$
b. $20 \cdot 5 \cdot 11 \cdot 19$
c. $2 \cdot 5 \cdot 5 \cdot 11 \cdot 19$
d. $2 \cdot 5 \cdot 10 \cdot 11 \cdot 19$
e. $2 \cdot 2 \cdot 5 \cdot 5 \cdot 11 \cdot 19$

24. An integer is abundant if its positive integer factors excluding the integer itself, have a sum that is greater than the integer. How many of the integers 12, 14, 16, and 18 are abundant?

PA 22

a. 0
b. 1
c. 2
d. 3
e. 4

25. The ratio of Moe's height to Ruby's height Is 4:5. The ratio of Ruby's height to Goldie's height is 7:6. What is the ratio of Moe's height to Goldie's height?

PA 25

a. 2:3
b. 5:6
c. 6:5
d. 14:15
e. 24:35

ALGEBRA 1

1. If $6(x - 10) = -11$, then $x = ?$

A1 01

   a. $-\frac{71}{6}$

   b. $-\frac{21}{6}$

   c. $-\frac{11}{6}$

   d. $\frac{49}{6}$

   e. $\frac{71}{6}$

2. What is the product of the solutions of the 2 equations below?

A1 03

$$10x = 15$$

$$14 + 3y = 41$$

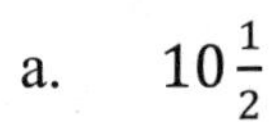

   a. $10\frac{1}{2}$

   b. 12

   c. $13\frac{1}{2}$

   d. $20\frac{1}{2}$

   e. $27\frac{1}{2}$

3. If $4x^2 + 32x = 80$, what are the possible values of x ?

A1 02

a. –20 and 2
b. –10 and 2
c. –2 and 10
d. –2 and 20
e. 8 and 12

4. When the vector $wi - 4j$ is added to the vector $5i + xj$, the sum is $-2i + 5j$. What are the values of w and x ?

A1 04

a. $w = -9$ and $x = 7$
b. $w = 9$ and $x = -7$
c. $w = -3$ and $x = 1$
d. $w = 3$ and $x = -1$
e. $w = -7$ and $x = 9$

5. If $11 - 4x = -21$, then $3x =$?

A1 05

a. -15
b. 8
c. 15
d. 24
e. 32

6. The solution set of which of the following equations is the set of real numbers that are 6 units from -2 ?

A1 08

a. $|x+2|=6$
b. $|x-2|=6$
c. $|x+6|=2$
d. $|x-6|=2$
e. $|x+6|=-2$

7. When $\frac{1}{5}x+\frac{2}{3}x=1$, what is the value of x ?

A1 07

a. $\frac{1}{13}$
b. $\frac{15}{13}$
c. $\frac{13}{3}$
d. 6
e. 15

A1 06

8. What is the value of $\log_3 81$?

a. 3
b. 4
c. 27
d. 78
e. 84

9. What is the real value of x in the equation $\log_3 54 - \log_3 2 = \log_4 x$?

A1 09

a. 2
b. 52
c. 64
d. 81
e. 108

10. In the figure below, all segments that meet do so at right angles. What is the area, in square units, of the shaded region?

A1 10

a. 6
b. 7
c. 8
d. $10\frac{1}{2}$
e. $12\frac{1}{2}$

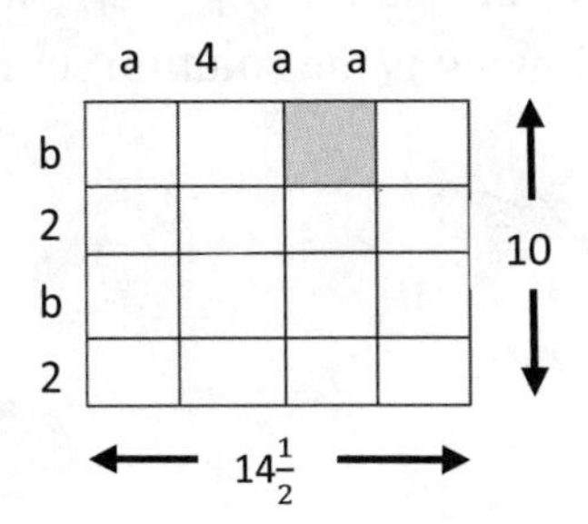

11. The perimeter of a certain scalene triangle is 120 inches. The side lengths of the triangle are represented by $x + 30$, $3x + 18$, and $5x$, respectively. What is the length, in inches, of the longest side of the triangle?

A1 11

a. 8
b. 38
c. 40
d. 42
e. 72

12. Given that $w\begin{bmatrix} 3 & 5 \\ 2 & 8 \end{bmatrix} = \begin{bmatrix} x & y \\ z & 28 \end{bmatrix}$ for some real number w, what is $x + y$?

A1 12

a. $\frac{35}{2}$
b. 14
c. 27
d. 28
e. 35

A1 14

13. If x and y can be any integers such that $y > 11$ and $3x + y = 21$, which of the following is the solution set for x ?

a. $x \leq 3$
b. $x \geq 3$
c. $x \leq 2$
d. $x \geq 2$
e. $x \leq 0$

14. The inequality $4(x + 7) < 5(x - 3)$ is equivalent to which of the following inequalities?

A1 15

a. $x < 13$
b. $x > 13$
c. $x < 23$
d. $x > 43$
e. $x < 43$

15. If p, q and r are positive integers such that $p^q = s$ and $r^q = t$, then $st = ?$

A1 13

a. pr^q
b. pr^{2q}
c. $(pr)^q$
d. $(pr)^{2q}$
e. $(pr)^{q^2}$

16. Given $p = xy^2$, $p = 6200$, and $y = 100$, what is x ?

A1 17

a. 0.62
b. 6.2
c. 31
d. 62
e. 310

17. Which of the following is the solution statement for the inequality shown below?

A1 18

$$-3 < 3 - 2x < 9$$

a. $-3 < x < 9$
b. $-3 < x$
c. $-3 < x < 3$
d. $3 < x$
e. $x < -3$ or $x > 3$

18. Which of the following number line graphs shows the solution set to the inequality $|x - 7| < -2$?

A1 16

a. 5 9

b. 5 9

c. 5 9

d. 5 9

e. 5 9
(empty set)

ALGEBRA 2

1. Reese purchased a car that had a purchase price of $17,800, which included all other costs and tax. He paid $800 as a down payment and got a loan for the rest of the purchase price. Reese paid off the loan by making 60 payments of $315 each. The total of all his payments, including the down payment, was how much more than the car's purchase price?

A2 01

   a. $1,100
   b. $1,900
   c. $17,000
   d. $18,900
   e. $19,700

2. Steve is trying to decide whether to buy a season pass to a local ski resort this winter. The cost of an individual daily lift ticket is $96.00, and the cost of a season pass is $1200.00. What is the minimum number of days Steve must ski at the resort this season in order for the cost of a season pass to be less than the total cost of buying an individual lift ticket for each day he skis there?

A2 02

   a. 10
   b. 11
   c. 12
   d. 13
   e. 14

3. Discount tickets to a volleyball tournament sell for \$6.00 each. David spent \$126.00 on discount tickets, \$52.50 less than if he had bought the tickets at the regular price. What was the regular ticket price?

A2 04

a. \$3.50
b. \$7.50
c. \$8.50
d. \$10.00
e. \$12.00

4. Vehicle A averages 18 miles per gallon of gasoline, and vehicle B averages 32 miles per gallon of gasoline. At these rates, how many more gallons of gasoline does vehicle A need than vehicle B to make a 1,152 mile trip?

A2 05

a. 25
b. 28
c. 36
d. 50
e. 64

5. On the first day of school, Mr. Catley gave his ninth grade students 8 new math problems to solve. On each day of school after that, he gave the students 4 new math problems to solve. In the first 30 days of school, how many math problems had he given the students to solve?

A2 03

a. 42
b. 124
c. 128
d. 132
e. 232

6. Mel earns her regular pay of \$12.50 per hour for up to 40 hours of work in a week. For each hour over 40 hours of work in a week, Mel is paid $1\frac{1}{2}$ times her regular pay. How much does Mel earn for a week in which she works 44 hours?

A2 06

a. \$506.00
b. \$550.00
c. \$575.00
d. \$600.00
e. \$825.25

7. A container is $\frac{1}{6}$ full of water. After 15 cups of water are added, the container is $\frac{11}{12}$ full. What is the volume of the container, in cups?

A2 07

a. $11\frac{1}{4}$
b. 20
c. 24
d. 26
e. $27\frac{1}{2}$

8. The total cost of renting a moving truck is $45.00 for each day the moving truck is rented plus $12\frac{1}{2}$¢ for each mile the moving truck is driven. What is the total cost of renting the moving truck for one week and driving it 410 miles?

A2 10

a. $321.25
b. $352.75
c. $364.20
d. $366.25
e. $544.00

9. In the figure below, the area of the larger square is 108 square centimeters, and the area of the smaller square is 12 square centimeters. What is x, in centimeters?

A2 09

a. 3
b. $2\sqrt{3}$
c. $4\sqrt{3}$
d. 10
e. 96

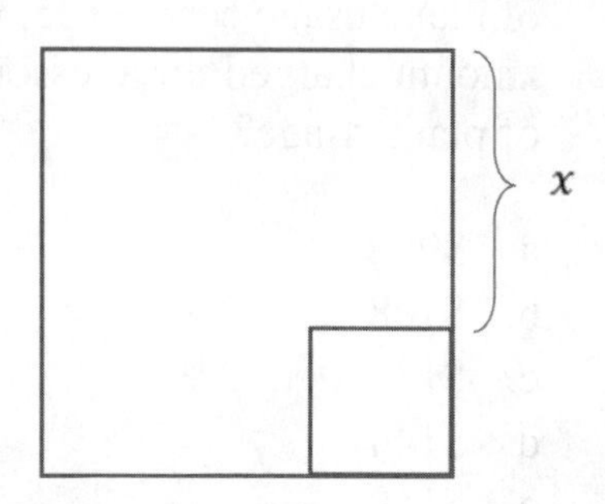

10. A printer prints 50 sheets of paper per minute. A second printer prints 75 sheets of paper per minute. The second printer starts printing paper 3 minutes after the first printer starts printing. Both printers stop printing 12 minutes after the first printer started. Together, the 2 printers print how many sheets of paper?

A2 08

a. 600
b. 675
c. 1275
d. 1350
e. 1725

11. A skydiving course charges \$75 per lesson, plus an additional fee for use of a plane. The charge for the use of a plane varies directly with the square root of the time the plane is used. If a lesson plus 25 minutes of plane usage costs \$105, what is the total amount charged for a lesson having 64 minutes of plane usage?

A2 11

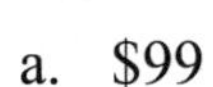

a. \$99
b. \$108
c. \$123
d. \$144
e. \$150

12. Tara cut a pipe 21 feet long into 2 pieces. The ratio of the lengths of the 2 pieces is 3:4. What is the length to the nearest foot, of the shorter piece?

A2 12

a. 3
b. 6
c. 7
d. 9
e. 12

13. An artist makes a profit of $(600x - x^2)$ dollars from selling x paintings. What is the fewest number of paintings the artist can sell to make profit of at least $50,000?

A2 14

a. 100
b. 200
c. 400
d. 500
e. 600

14. The sum of 2 positive numbers is 171. The lesser number is 15 more than the square root of the greater number. What is the value of the greater number minus the lesser number?

A2 13

a. 15
b. 105
c. 117
d. 144
e. 167

ALGEBRA 3

1. For $i^2 = -1, (i-5)^2 = ?$

   a. 24
   b. 26
   c. $24 - 5i$
   d. $24 - 10i$
   e. $25 - 5i$

2. Which of the following expressions is a factor of $x^3 - 27$?

   a. $x - 3$
   b. $x + 3$
   c. $x + 27$
   d. $x^2 + 9$
   e. $x^2 - 3x + 9$

3. Which of the following expressions is equivalent to $\frac{1}{3}x^2\,(9y - 6x + 3y + 6x)$?

A3 01

   a. $4x^2y$
   b. $8xy$
   c. $2x^2y + 12y$
   d. $4x^2y - 4x^3$
   e. $2x^2y + 12y - x^3 + 6x$

4. Which of the following is equivalent to $(5r^4)^3$?

A3 04

a. $125r^7$
b. $125r^{12}$
c. $15r^{12}$
d. $15r^7$
e. $5r^{12}$

5. Which of the following is a factored form of the expression $4x^2 - 17x - 15$?

A3 03

a. $(x-5)(4x+3)$
b. $(x-3)(4x-5)$
c. $(x-3)(4x+5)$
d. $(x+3)(4x-5)$
e. $(x+5)(4x-3)$

6. The expression $-9x^2(7x^4 - 6x^5)$ is equivalent to:

A3 08

a. $-63x^6 + 54x^7$
b. $-63x^6 - 54x^7$
c. $-63x^8 + 54x^{10}$
d. $-63x^8 - 54x^{10}$
e. $-9x^3$

7. $7x^2 - 44x - 70x^2 + 51 + 45x$ is equivalent to:

A3 09

a. $-11x^2$
b. $-11x^6$
c. $-63x^4 + x^2 + 51$
d. $-63x^2 + x + 51$
e. $77x^2 + x + 51$

8. The expression $(5y + 6)(y - 4)$ is equivalent to:

A3 06

a. $5y^2 - 10$
b. $5y^2 - 24$
c. $5y^2 + 26y - 24$
d. $5y^2 - 14y - 24$
e. $5y^2 + 14y - 24$

9. Which of the following mathematical expressions is equivalent to the verbal expression "A number, when cubed is equal to 16 less than the product of itself and 20"?

A3 07

a. $3x = 20x - 16$
b. $3x = 20x - 16x$
c. $x^3 = 20x + 16$
d. $x^3 = 16 - 20x$
e. $x^3 = 20x - 16$

10. For all $x > 1$, the expression $\frac{6x^5}{6x^2}$ equals:

A3 10

a. $\frac{1}{3}$
b. $-x^3$
c. x^3
d. $-\frac{1}{x^3}$
e. $\frac{1}{x^3}$

11. For all real numbers p and q such that the product of q and 5 is p, which of the following expressions represents the sum of q and 5 in terms of p?

A3 12

a. $p + 5$
b. $5p + 5$
c. $5(p + 5)$
d. $\frac{p+5}{5}$
e. $\frac{p}{5} + 5$

12. Danica went into an electronics store to price video games. All video games were discounted 15% off the marked price. Danica wanted to program her calculator so she could input the marked price and the discounted price would be the output. Which of the following is an expression for the discounted price on a marked price of p dollars?

A3 11

a. $p - 0.15p$
b. $p - 0.15$
c. $p - 15p$
d. $p - 15$
e. $0.15p$

13. The fixed cost of manufacturing shoes in a factory are \$1250.00 per day. The variable costs are \$1.75 per shoe. Which of the following expressions can be used to model the cost of manufacturing x shoes in one day?

A3 13

a. $\$1401.75x$
b. $\$1.75x - \$1250.00$
c. $\$1250.00x + \$1.75$
d. $\$1250.00 - \$1.75x$
e. $\$1250.00 + \$1.75x$

14. Sailor Sam has decided to start a company that will produce and sell houseboats. In order to begin this venture, he must invest $8 million in a houseboat production plant. The cost to produce each houseboat will be $24,000, and the selling price will be $95,000. Accounting for the cost of the production plant, which of the following expression represents the profit, in dollars, that Sailor Sam will realize when x houseboats are produced and sold?

A3 14

a. $71{,}000x - 8{,}000{,}000$
b. $119{,}000x - 8{,}000{,}000$
c. $7{,}881{,}000x$
d. $95{,}000x$
e. $71{,}000x$

ALGEBRA 4

1. The 3 statements below are true for the elements of sets W, X, Y and Z.

A4 03

I. All elements of W are elements of X.
II. All elements of Y are elements of W.
III. No elements of Z are elements of X.

Which of the following statements *must* be true?

a. All elements of W are elements of Z.
b. All elements of X are elements of Z.
c. All elements of X are elements of Y.
d. All elements of Y are elements of Z.
e. All elements of Y are elements of X.

2. If x and y are real numbers such that $x < 0$ and $y > 0$, then which of the following is equivalent to $|x| + |y|$?

A4 04

a. $|x + y|$
b. $|y - x|$
c. $|y| - |x|$
d. $x + y$
e. $y - x$

3. If $4^x = 125$, then which of the following must be true?

A4 01

a. $1 < x < 2$
b. $2 < x < 3$
c. $3 < x < 4$
d. $4 < x < 5$
e. $5 < x$

4. If $x > y$ and $y > 5$, then what is the smallest possible integer value of $x + y$?

A4 02

a. 5
b. 6
c. 10
d. 11
e. 12

5. Which of the following expressions has a negative value for all a and b such that $a < 0$ and $b > 0$?

A4 07

a. $b - a$
b. $a + b$
c. $(ab)^4$
d. a^3b
e. $\frac{a^2}{b}$

6. If $x < -1$, then which of the following has the GREATEST value?

A4 06

a. $\frac{x}{2}$
b. $\sqrt{x \cdot x}$
c. $x \cdot x$
d. $x \cdot x \cdot x$
e. $4x$

7. For which of the following conditions will the product of integers x and y *always* be an odd integer?

A4 05

a. x is an odd integer.
b. y is an odd integer.
c. x and y are both odd integers.
d. x and y are both even integers.
e. x is an odd integer and y is an even integer.

GEOMETRY 1

1. In the standard (x, y) coordinate plane, $A(6, 3)$ lies on a circle with center $(3, -1)$ and radius 5 coordinate units. What are the coordinates of the image of A after the circle is rotated 90° clockwise about the center of the circle?

G1 01

   a. $(-1, 3)$
   b. $(3, -1)$
   c. $(6, -5)$
   d. $(7, -4)$
   e. $(8, -1)$

2. The sides of a square are 5 cm long. One vertex of the square is at $(-2, 0)$ on a square coordinate grid marked in centimeter units. Which of the following points could also be a vertex of a square?

G1 04

   a. $(-7, 0)$
   b. $(4, 3)$
   c. $(-1, -1)$
   d. $(2, 0)$
   e. $(0, -2)$

3. A particular circle in the standard (x, y) coordinate plane has an equation of $x^2 + (y - 4)^2 = 14$. What is the radius of the circle, in coordinate units, and the coordinates of the center of the circle?

G1 03

	radius	center
a.	$\sqrt{14}$	$(0, 4)$
b.	7	$(0, 4)$
c.	14	$(0, 4)$
d.	$\sqrt{14}$	$(0, -4)$
e.	7	$(0, -4)$

4. A circle in the standard (x, y) coordinate plane has center $C(3, -1)$ and passes through $A(6, -5)$. Line segment $\overline{AB}$ is a diameter of this circle. What are the coordinates of point B ?

G1 02

a. $(-6, 5)$
b. $(-1, 2)$
c. $(0, 3)$
d. $(5, 10)$
e. $(8, -1)$

5. The equations below are equations of a system where $a, b,$ and c are positive integers.

G1 05

$$ay - bx = c$$

$$-2(bx - ay) = 2c$$

Which of the following describes the graph of at least 1 such system of equations in the standard (x, y) coordinate plane?

I. 2 parallel lines
II. 2 intersecting lines
III. A single line

a. I only
b. II only
c. III only
d. I or II only
e. I, II, or III

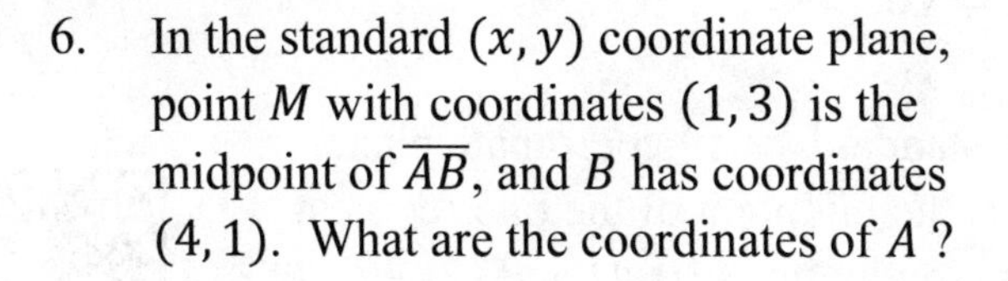

6. In the standard (x, y) coordinate plane, point M with coordinates $(1, 3)$ is the midpoint of $\overline{AB}$, and B has coordinates $(4, 1)$. What are the coordinates of A ?

G1 06

a. $(5, 4)$
b. $(7, -1)$
c. $(2.5, 2)$
d. $(-2, 5)$
e. $(2, -5)$

7. In the standard (x, y) coordinate plane, the midpoint of $\overline{AB}$ is $(-2, 5)$ and A is located at $(-6, 2)$. If (x, y) are the coordinates of B, what is the value of $x + y$?

G1 07

a. –5
b. –3
c. 6
d. 10
e. 16

8. Consider the graph of the equation $y = \frac{3x - 18}{2x - 10}$ in the standard (x, y) coordinate plane. Which of the following equations represents the *vertical* asymptote of the graph?

G1 08

a. $x = 2$
b. $x = 5$
c. $x = 6$
d. $x = 8$
e. $x = 10$

9. In the standard (x, y) coordinate plane, what is the midpoint of the line segment that has endpoints $(10, 4)$ and $(2, -2)$?

G1 09

a. $(-8, -6)$
b. $(6, 1)$
c. $(7, 0)$
d. $(8, 6)$
e. $(12, -6)$

10. What is the slope of the line through $(5,-4)$ and $(9,5)$ in the standard (x,y) coordinate plane?

G1 10

a. 1

b. 9

c. –9

d. $\frac{9}{4}$

e. $-\frac{9}{4}$

11. Lines f and g lie in the standard (x,y) coordinate plane. An equation for line f is $y = 0.81x - 50$. The slope of line g is 0.2 greater than the slope of line f. What is the slope of line g ?

G1 12

a. 0.162
b. 0.83
c. 1.01
d. 4.05
e. 5

12. The slope of the line with equation $y = ax + b$ is less than the slope of the line with equation $y = cx + b$. Which of the following statements *must* be true about the relationship between a and c ?

G1 11

a. $a \leq c - 1$
b. $a < c$
c. $a = c$
d. $a > c$
e. $a \geq c$

13. In the standard (x, y) coordinate plane, what is the slope of the line $3x + 8y = 5$?

G1 13

a. $-\frac{3}{8}$

b. $\frac{3}{5}$

c. –3

d. 3

e. 5

14. What is the slope of any line perpendicular to the line $7x + 5y = 10$ in the standard (x, y) coordinate plane?

G1 14

a. -7

b. $-\frac{7}{5}$

c. $\frac{5}{7}$

d. $-\frac{5}{7}$

e. 7

15. For some real number m, the graph of the line $y = (m - 2)x + 4$ in the standard (x, y) coordinate plane passes through $(-3, -2)$. What is the slope of this line?

G1 15

a. -4
b. -3
c. 2
d. 3
e. 4

16. Which of the following is the slope of a line parallel to the line $y = \frac{5}{3}x - 8$ in the standard (x, y) coordinate plane?

G1 18

a. –8

b. $-\frac{3}{5}$

c. 5

d. $\frac{3}{5}$

e. $\frac{5}{3}$

17. The points $A(2, 3)$ and $B(11, 9)$ lie in the standard (x, y) coordinate plane. Point C lies on $\overline{AB}$ between A and B such that the length of $\overline{AB}$ is 3 times the length of $\overline{CA}$. What are the coordinates of C ?

G1 16

a. $(4, 6)$
b. $(5, 5)$
c. $(6, 6)$
d. $(7, 8)$
e. $(8, 7)$

18. Rectangle $ABCD$ has vertices $A(4,4)$, $B(-2,2)$, and $C(2,-6)$. These vertices are graphed below in the standard (x,y) coordinate plane. What are the coordinates of vertex D ?

G1 17

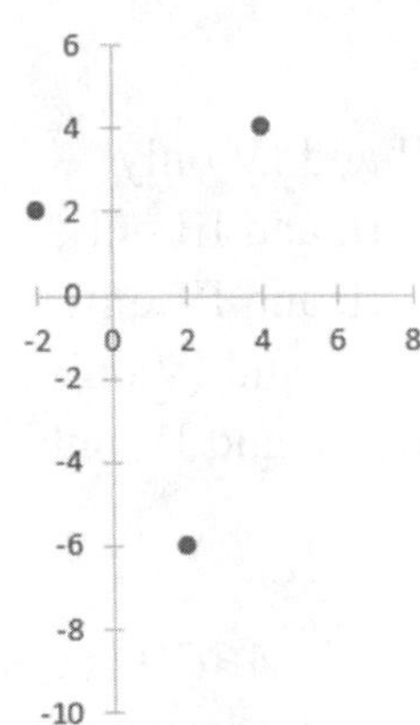

a. $(9,-5)$
b. $(8,-4)$
c. $(7,-3)$
d. $(6,-3)$
e. $(0,-10)$

19. The point $(-2,-3)$ is shown in the standard (x,y) coordinate plane below. Which of the following is another point on the line through the point $(-2,-3)$ with a slope of $-\frac{1}{3}$?

G1 19

a. $A(-5,-4)$
b. $B(-3,0)$
c. $C(-1,-6)$
d. $D(1,-4)$
e. $E(1,-2)$

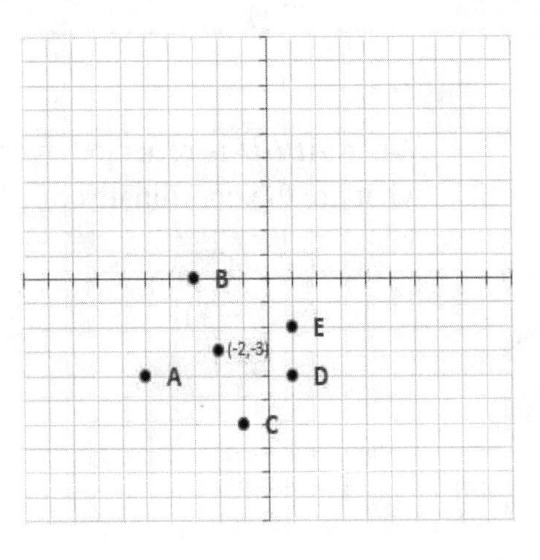

20. What are the quadrants of the standard (x, y) coordinate plane below that contain points on the graph of the equation $6x + 3y = -9$?

G1 20

a. II and IV only
b. I, II, and III only
c. I, II, and IV only
d. I, III, and IV only
e. II, III, and IV only

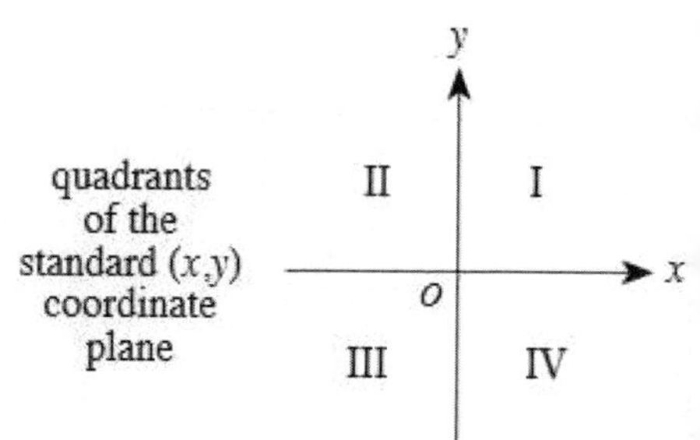

21. Trapezoid $ABCD$ is graphed in the standard (x, y) coordinate plane below.

G1 21

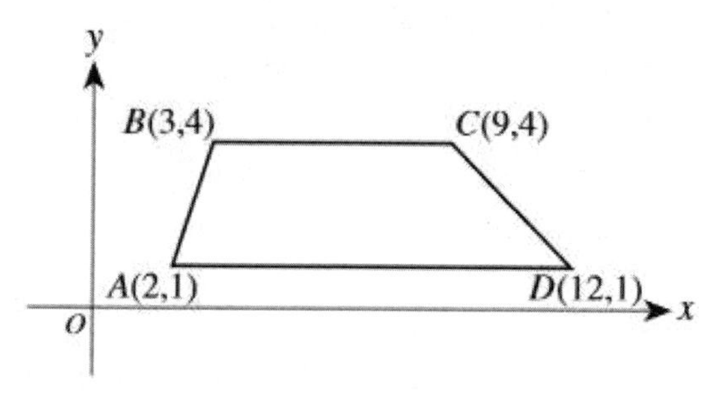

When $ABCD$ is reflected over the x-axis to $A'B'C'D'$, what are the coordinates of B' ?

a. $(3, -4)$
b. $(-3, -4)$
c. $(-3, 4)$
d. $(4, 3)$
e. $(4, -3)$

G1 22

22. The graph of the equations $y = x + 2$ and $y = (x + 2)^2$ are shown in the standard (x, y) coordinate plane below. What real values of x, if any, satisfy the inequality $(x + 2)^2 < (x + 2)$?

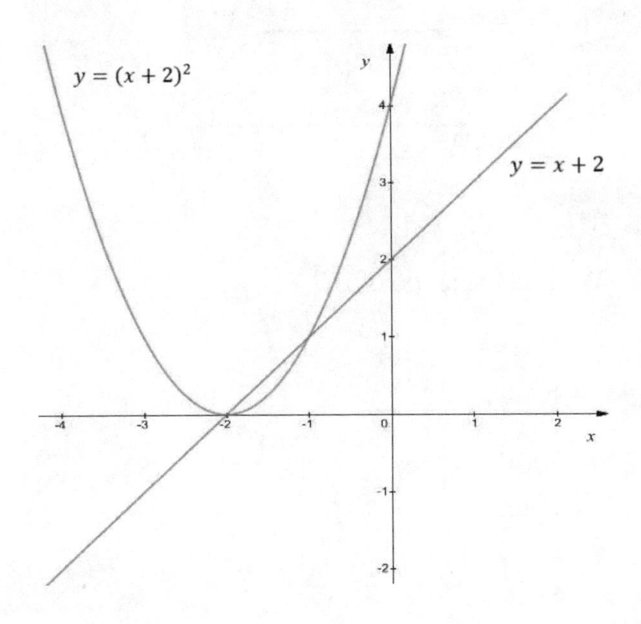

a. No real values
b. $x < 0$ and $x > 1$
c. $x < -2$ and $x > -1$
d. $0 < x < 1$
e. $-2 < x < -1$

23. Trapezoid $ABCD$ is graphed in the standard (x, y) coordinate plane below.

G1 23

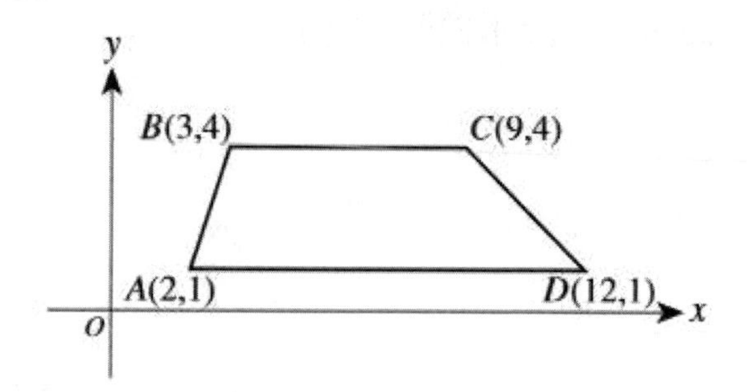

What is the slope of $\overline{AB}$?

a. –3

b. –1

c. 3

d. $\frac{1}{3}$

e. $\frac{4}{5}$

GEOMETRY 2

1. In the parallelogram $PQRS$ below, $\overline{SQ}$ is a diagonal, the measure of $\angle QRS$ is 96°, and the measure of $\angle PQS$ is 49°. What is the measure of $\angle PSQ$?

G2 02

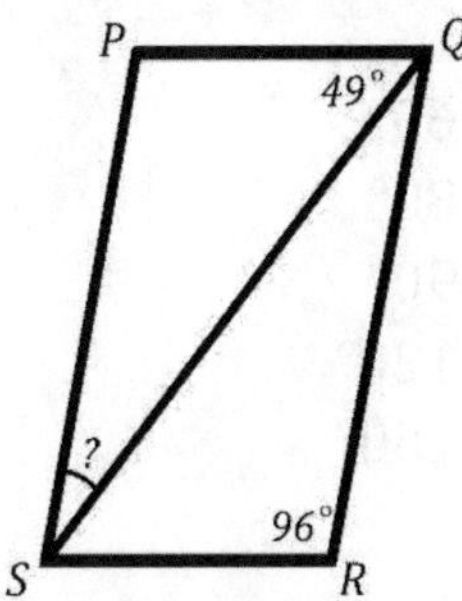

a. 25°
b. 35°
c. 37°
d. 48°
e. 145°

2. In ΔPQR, the sum of the measures of $\angle P$ and $\angle Q$ is 62°. What is the measure of $\angle R$?

G2 03

a. 28°
b. 62°
c. 118°
d. 124°
e. 128°

3. Ray $\overrightarrow{BP}$ bisects $\angle ABC$, the measure of $\angle ABC$ is $6x°$ and the measure of $\angle ABP$ is $(2x + 16)°$. What is the measure of $\angle PBC$?

G2 05

a. 16°
b. 32°
c. 42°
d. 48°
e. 96°

4. A teacher drew the circle graph below describing his time spent working at school in 1 day. His principal said that the numbers of hours listed were correct, but that the central angle measures for the sectors were not correct. What should be the central angle measures for the meetings sector?

G2 04

a. $60°$
b. $80°$
c. $90°$
d. $120°$
e. $160°$

5. Lines a, b, c and d are shown below and $c \parallel d$. Which of the following is the set of all angles that *must* be supplementary to $\angle x$?

G2 01

a. $\{1, 3\}$
b. $\{1, 3, 4, 7\}$
c. $\{1, 3, 8, 11\}$
d. $\{1, 3, 4, 7, 8, 11\}$
e. $\{1, 3, 4, 7, 8, 11, 12, 15\}$

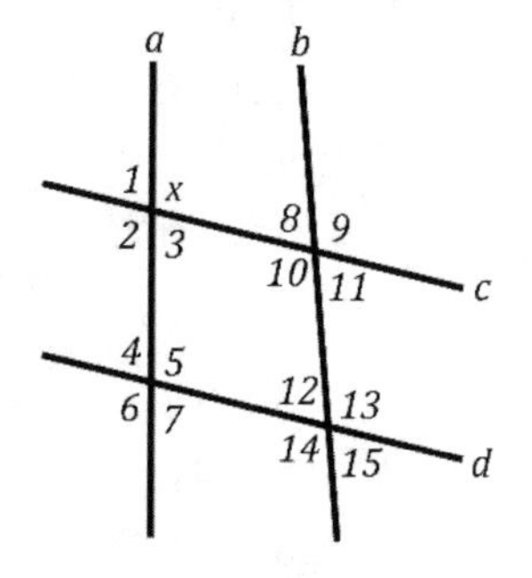

6. In the figure below, $L, N, O,$ and P are collinear. If $\angle OPM$ measures 52°, $\angle OMP$ measures 42°, and $\angle MNL$ measures 138°, what is the degree measure of $\angle OMN$?

G2 11

a. 42°
b. 44°
c. 45°
d. 52°
e. 86°

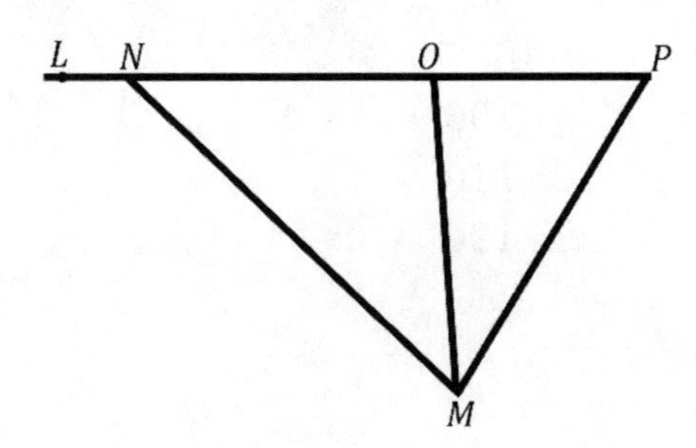

7. Given the triangle shown below with exterior angles that measure $x°, y°,$ and $z°$ as shown, what is the sum of $x, y,$ and z ?

G2 06

a. 180
b. 235
c. 305
d. 360
e. Cannot be determined

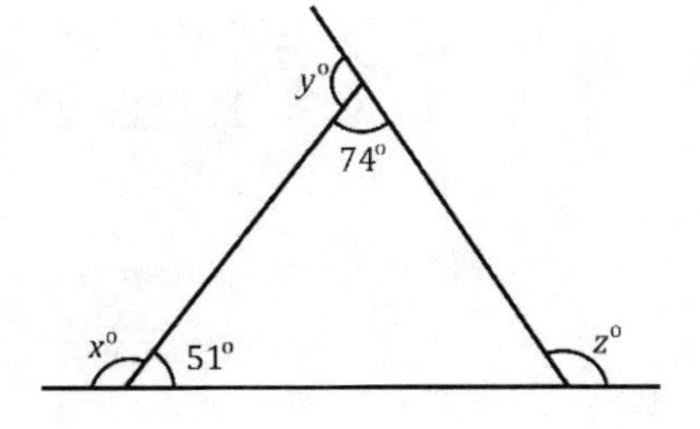

8. In the figure below, 2 nonadjacent sides of a regular nonagon (9 congruent sides and 9 congruent interior angles) are extended until they meet at point X. What is the measure of $\angle X$?

G2 09

a. 40°
b. 80°
c. 100°
d. 110°
e. 120°

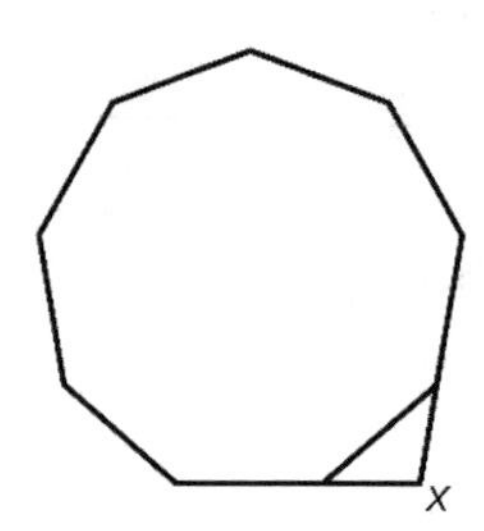

9. For trapezoid $ABCD$ shown below, $\overline{AB} \parallel \overline{DC}$, the measures of the interior angles are distinct, the measure of $\angle B$ is $x°$ and the measure of $\angle C$ is $y°$. Which equation below demonstrates the relation between these two angles?

G2 08

a. $y° = x°$
b. $y° = 90° + x°$
c. $y° = 90° - x°$
d. $y° = 180° + x°$
e. $y° = 180° - x°$

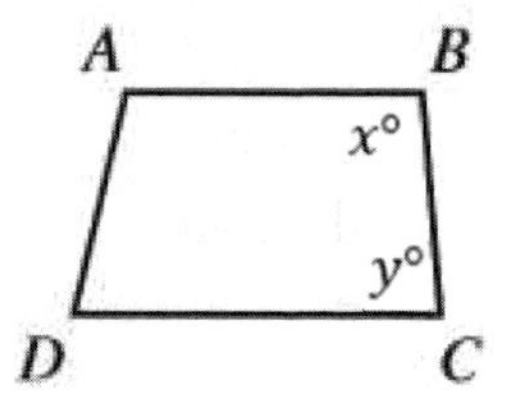

10. In the figure below, $\Delta ABC \cong \Delta EBD$. Which of the following congruences is NOT necessarily true?

a. $\angle CBA \cong \angle DBE$
b. $\angle ACB \cong \angle EDB$
c. $\overline{AB} \cong \overline{CB}$
d. $\overline{AC} \cong \overline{ED}$
e. $\overline{AB} \cong \overline{EB}$

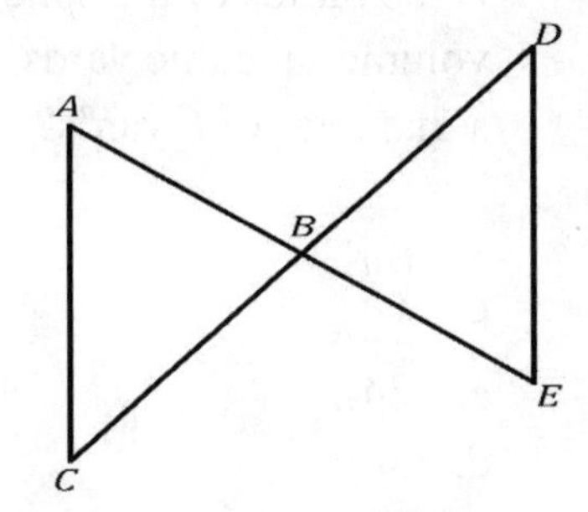

11. In the circle shown below, chords $\overline{DB}$ and $\overline{AC}$ intersect at O, which is the center of the circle, and the measure of $\angle OCD$ is 30°. What is the degree measure of minor arc $\widehat{BC}$?

a. 25°
b. 50°
c. 75°
d. 100°
e. Cannot be determined

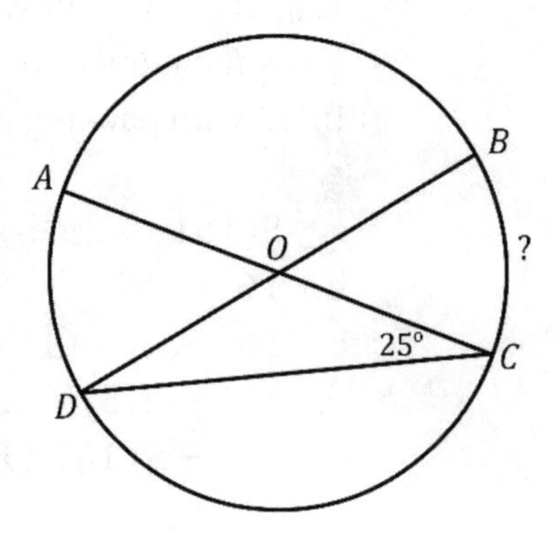

GEOMETRY 3

1. The volume of a sphere is $\frac{4\pi r^3}{3}$, where r is the radius of the sphere. What is the volume, in cubic yards, of a sphere with a diameter of 6 yards?

G3 01

   a. 8π
   b. 12π
   c. 24π
   d. 36π
   e. 288π

2. A cube shaped box has an interior side length of 16 inches and is used to ship a right circular cylinder with a diameter of 10 inches and a height of 11 inches. The interior of the box not occupied by the cylinder is filled with packing material. Which of the following numerical expressions gives the number of cubic inches of the box filled with packing material?

G3 03

   a. $5(16)^2 - 2\pi(5)(11)$
   b. $16^3 - \pi(11)^3$
   c. $16^3 - \pi(5)(11)^2$
   d. $16^3 - \pi(5)^2(11)$
   e. $16^3 - \pi(10)^2(11)$

3. You completely submerge a solid toy in a rectangular tank that has a base of 60 centimeters by 40 centimeters and is filled with water to a depth of 25 centimeters. The toy sinks to the bottom, and the water level goes up 0.75 centimeters. What is the volume, in cubic centimeters, of the toy?

G3 02

a. 1800
b. 1125
c. 1000
d. 900
e. 750

4. Cube A has an edge length of 3 inches. Cube B has an edge length triple that of cube A. What is the volume, in cubic inches, of cube B?

G3 05

a. 9
b. 27
c. 81
d. 243
e. 729

5. What is the surface area, in square inches, of a 12 inch cube?

G3 04

a. 1728
b. 864
c. 720
d. 576
e. 432

6. A room has a rectangular floor that is 18 feet by 27 feet. What is the area of the floor in square *yards* ?

G3 07

a. 30
b. 45
c. 54
d. 84
e. 162

7. The diameter of one circle is 14 inches long. The diameter of a second circle is 25% shorter than the diameter of the first circle. To the nearest square inch, how much smaller is the area of the second circle than the area of the first circle?

G3 08

a. 6
b. 67
c. 87
d. 154
e. 269

8. Adam calculates that he needs 540 square feet of new carpet. But the type of carpet that he wants is priced by the square *yard.* How many square yards of carpet does he need?

G3 09

a. 25
b. 60
c. 90
d. 135
e. 180

9. The edges of a cube are each 5 feet long. What is the surface area, in square feet, of this cube?

G3 06

   a. 25
   b. 30
   c. 75
   d. 125
   e. 150

10. Ken is making a tablecloth for a circular table 5 feet in diameter. The finished tablecloth needs to hang down 4 inches over the edge of the table all the way around. To finish the edge of the tablecloth, Ken will fold under and sew down 1 inch of the material all around the edge. Ken is going to use a single piece of rectangular fabric that is 80 inches wide. What is the shortest length of fabric, in inches, Ken could use to make the fabric tablecloth without putting any separate pieces of fabric together?

G3 10

   a. 20
   b. 35
   c. 50
   d. 65
   e. 70

11. Nathan plans to paint the 4 walls of his bedroom with 2 coats of paint. The walls are rectangular, and each wall is 10 feet by 12 feet. He will not need to paint the single 4-foot-by-6-foot rectangular window in his room and the 4-foot-by-8.5-foot rectangular door. Nathan knows that each gallon of paint covers between 350 and 400 square feet. Which of the following is the minimum number of 1-gallon cans of paint Nathan need to buy to paint his bedroom?

G3 11

a. 1
b. 2
c. 3
d. 4
e. 5

12. The perimeter of a parallelogram is 80 inches, and 1 side measures 16 inches. What are the lengths, in inches, of the other 3 sides?

G3 12

a. 16, 16, 32
b. 16, 20, 20
c. 16, 24, 24
d. 16, 32, 32
e. Cannot be determined

13. The length of a rectangle with an area of 108 square centimeters is 12 centimeters. What is the perimeter of the rectangle, in centimeters?

G3 14

a. 9
b. 18
c. 21
d. 33
e. 42

14. TV screen sizes are the diagonal length of the rectangular screen. Harry recently changed from watching a television with a 27-inch screen to a television with a 43-inch screen. If a building appeared 20 inches tall on the 27-inch screen, how tall, to the nearest inch, will it appear on the 43-inch screen?

G3 21

a. 22
b. 32
c. 36
d. 38
e. 40

15. The lengths of the corresponding sides of 2 similar right triangles are in the ratio 5:8. If the hypotenuse of the smaller triangle is 8 inches long, how many inches long is the hypotenuse of the larger triangle?

G3 13

a. 5
b. 12.8
c. 13
d. 16
e. 20

16. On an old road map, $\frac{1}{2}$ inch represents 26 miles. About how many miles apart are 2 towns that are $4\frac{1}{2}$ inches apart on this map?

G3 15

a. 26
b. 52
c. 78
d. 117
e. 234

17. In the figure below, the vertices of ΔPQR have (x, y) coordinates $(2,5)$, $(8,5)$, and $(3,9)$, respectively. What is the area of ΔPQR ?

G3 16

a. 12
b. $12\sqrt{2}$
c. $12\sqrt{3}$
d. 24
e. $24\sqrt{2}$

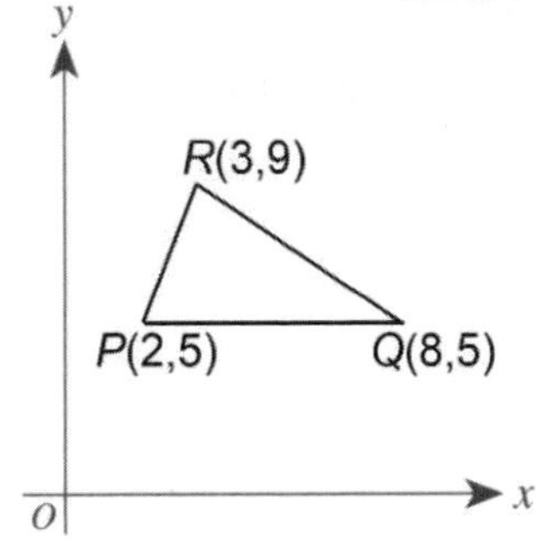

18. Jono is constructing the kite shown below. The kite includes 2 perpendicular supports, one length 36 inches and the other length 26 inches. The ends of the supports are connected with string to form a 4-sided figure that is symmetric with respect to the longer support. A layer of paper will cover the interior of the kite. Which of the following is closest to the area, in square inches, that Jono will need to cover with paper?

G3 18

a. 124
b. 234
c. 468
d. 702
e. 819

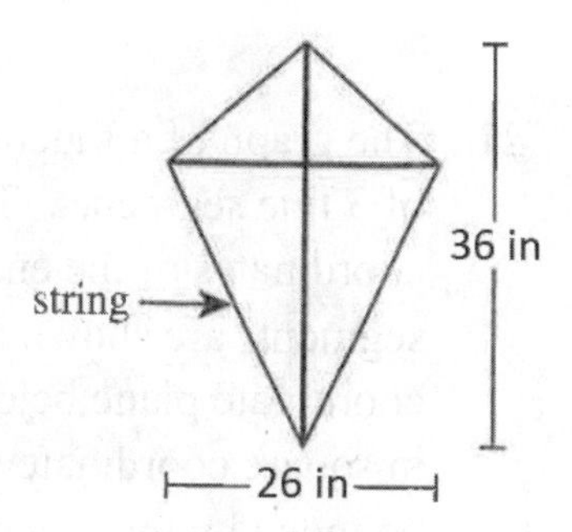

19. In the figure shown below, $A, B,$ and D lie on a circle whose center is O, a diameter is $\overline{AB}$, $\overline{CD}$ is perpendicular to $\overline{AB}$ at C, the length of $\overline{AD}$ is 8m, and the length of $\overline{BD}$ is 15m. What is the length, in meters, of $\overline{CD}$?

G3 17

a. $\frac{120}{17}$
b. $\frac{136}{15}$
c. 17
d. $\frac{255}{2}$
e. 120

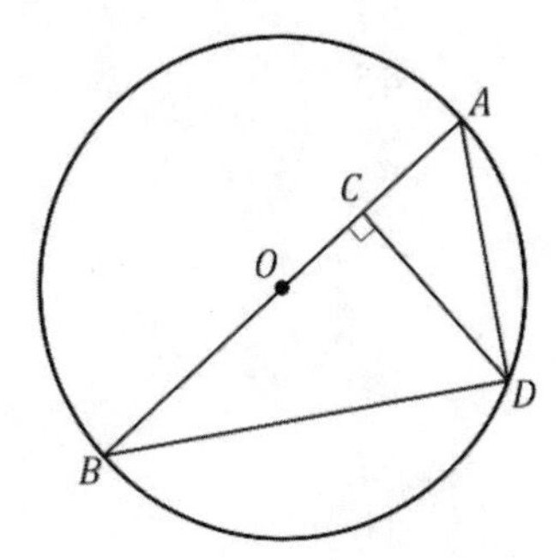

20. What is the surface area, in square inches, of a cube that has a volume of 64 cubic inches?

G3 19

a. 32
b. 48
c. 64
d. 96
e. 192

21. The graph of a function $y = f(x)$ consists of 3 line segments. The graph and the coordinates of the endpoints of the 3 line segments are shown in the standard (x, y) coordinate plane below. What is the area, in square coordinate units, of the region bounded by the graph of $y = f(x)$, the positive y-axis, and the positive x-axis ?

G3 20

a. 22.5
b. 27.5
c. 32.5
d. 37
e. 45

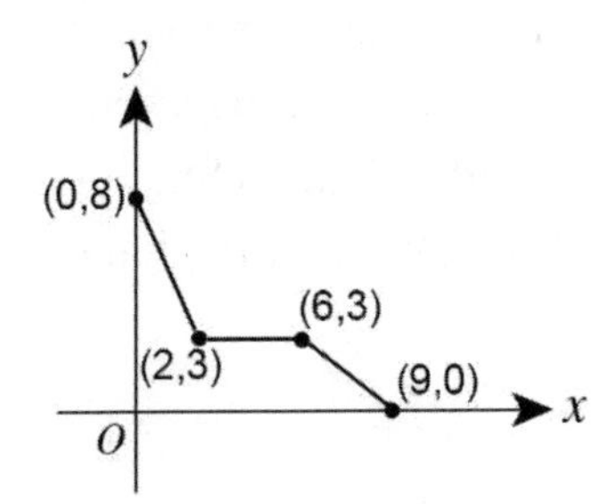

22. A swimming pool is a right circular cylinder with a diameter of 18 feet and a height of 5 feet, as seen in the diagram below. To the nearest cubic foot, what is the volume of water that will be in the pool when it is filled with water to a depth of 4 feet?

G3 23

(Note: The volume of a cylinder is given by $\pi r^2 h$, where r is the radius and h is the height.)

a. 226
b. 452
c. 1018
d. 1272
e. 4072

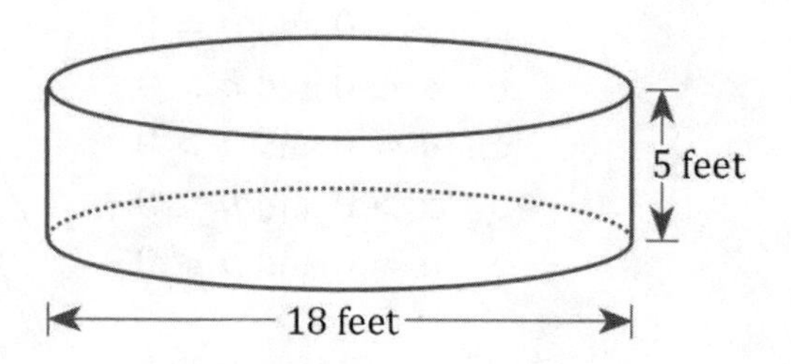

23. A hockey rink is being designed by combining a rectangle and 2 semicircles with the dimensions seen below. What is the perimeter, in feet, of the rink?

G3 22

a. $60 + 25\pi$
b. $120 + 25\pi$
c. $120 + 50\pi$
d. $170 + 25\pi$
e. $170 + 50\pi$

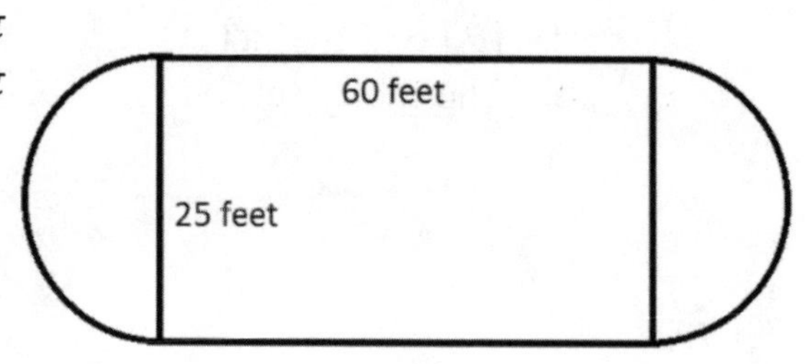

GEOMETRY 4

1. The functions $y = \cos x$ and $y = \cos(x + a) + b$, for constants a and b, are graphed in the standard (x, y) coordinate plane below. Which statement below about the values of a and b are true?

G4 03

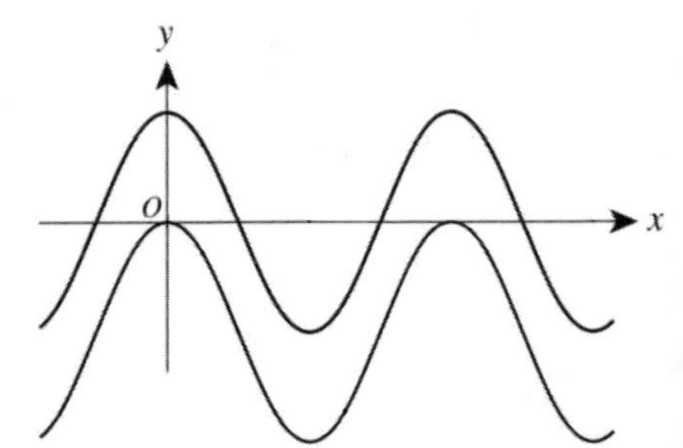

a. $a < 0$ and $b = 0$
b. $a < 0$ and $b > 0$
c. $a = 0$ and $b > 0$
d. $a > 0$ and $b < 0$
e. $a = 0$ and $b < 0$

2. For $0 \leq \theta \leq \pi$, $|\cos\theta| \geq 1$ is true for all and only the values of θ in which of the following sets?

G4 01

a. $\{0, \pi\}$
b. $\{0\}$
c. $\{\theta \mid 0 < \theta < \pi\}$
d. $\{\theta \mid 0 \leq \theta \leq \pi\}$
e. The empty set

3. The sides of an acute triangle measure 12 cm, 20 cm, and 24 cm, respectively. Which of the following equations, when solved for θ, give the measure of the smallest angle of the triangle?

G4 05

(Note: For any triangle with sided of length $a, b,$ and c that are opposite angles $A, B,$ and $C,$ respectively, $\frac{\sin A}{a} = \frac{\sin B}{b} = \frac{\sin C}{c}$ and $c^2 = a^2 + b^2 - 2ab\ cos\ C$.)

a. $\frac{\sin\theta}{12} = \frac{1}{20}$
b. $\frac{\sin\theta}{12} = \frac{1}{24}$
c. $\frac{\sin\theta}{24} = \frac{1}{12}$
d. $12^2 = 20^2 + 24^2 - 2(20)(24)\cos\theta$
e. $24^2 = 12^2 + 20^2 - 2(12)(20)\cos\theta$

4. For x such that $0 < x < \frac{\pi}{2}$, the expression $\frac{\sqrt{1-sin^2x}}{\cos x} - \sec x \cos x$ is equivalent to:

G4 02

a. 0
b. 1
c. 2
d. $-\tan x$
e. $\sin 2x$

5. For all real values of x, which of the following equations is true?

G4 04

a. $\sin(4x) + \cos(4x) = 4$
b. $\sin(4x) + \cos(4x) = 1$
c. $4\sin(4x) + 4\cos(4x) = 8$
d. $\sin^2(4x) + \cos^2(4x) = 4$
e. $\sin^2(4x) + \cos^2(4x) = 1$

6. The Statue of Liberty has a height of 93 meters. At a certain location, the angle of elevation formed by the level ground and the line from that location to the top of the Statue of Liberty is 60°. Which of the following expressions is equal to the distance, in meters, between that location and the center of the base of the statue?

G4 07

a. $93\cos 60°$

b. $93\sin 60°$

c. $93\tan 60°$

d. $\frac{93}{\sin 60°}$

e. $\frac{93}{\tan 60°}$

7. In the right triangle below, $0 < b < a$. One of the angle measures in the triangle is $tan^{-1}\left(\frac{b}{a}\right)$. What is $\sin\left[tan^{-1}\left(\frac{b}{a}\right)\right]$?

G4 06

a. $\frac{a}{b}$

b. $\frac{b}{a}$

c. $\frac{a}{\sqrt{a^2+b^2}}$

d. $\frac{b}{\sqrt{a^2+b^2}}$

e. $\frac{\sqrt{a^2+b^2}}{b}$

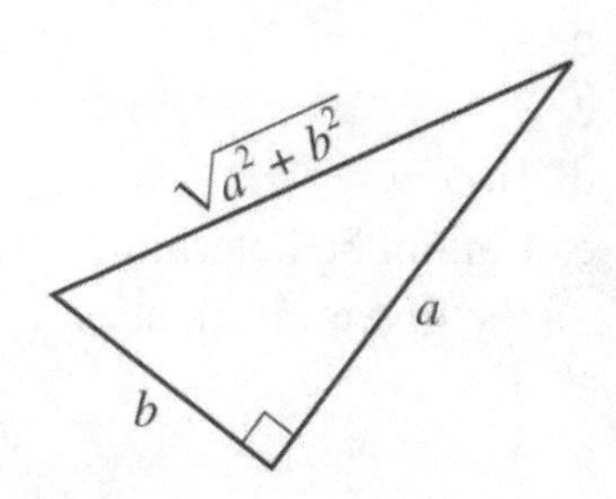

8. A line through the origin and $(14, 6)$ is shown in the standard (x, y) coordinate plane below. The angle between the line and the positive x-axis has measure θ. What is the value of $\tan\theta$?

G4 08

a. $\frac{\sqrt{58}}{3}$

b. $\frac{3}{\sqrt{58}}$

c. $\frac{7}{\sqrt{58}}$

d. $\frac{3}{7}$

e. $\frac{7}{3}$

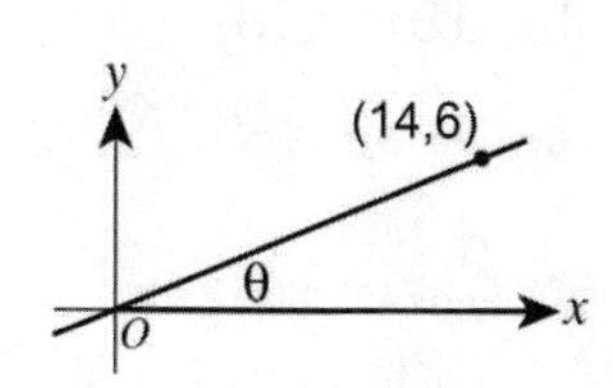

9. In ΔABD shown below, C is on $\overline{BD}$, the length of $\overline{AD}$ is 14 inches, and $\sin d = 0.75$. How many inches long is $\overline{CD}$?

G4 09

a. 3.8
b. 4.7
c. 9.3
d. 10.5
e. Cannot be determined from the given information

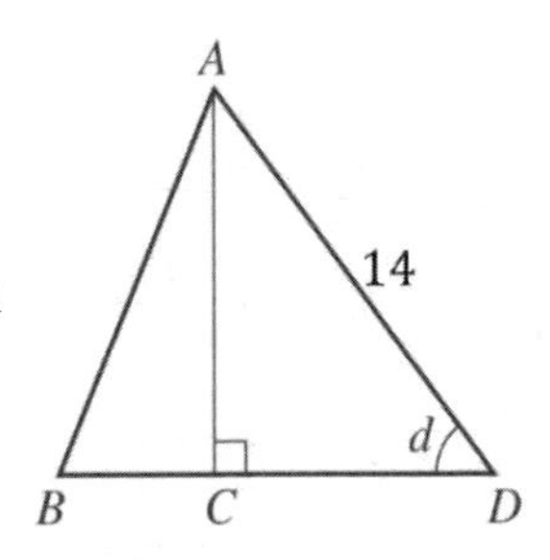

10. A boat coming towards a dock forms a triangle with a nearby lighthouse. According to the measurements given in the figure below, which of the following expressions gives the distance, in miles, from the boat to the dock?

G4 10

a. $18 \tan 48°$

b. $18 \cos 48°$

c. $18 \sin 48°$

d. $\frac{18}{\cos 48°}$

e. $\frac{30}{\sin 48°}$

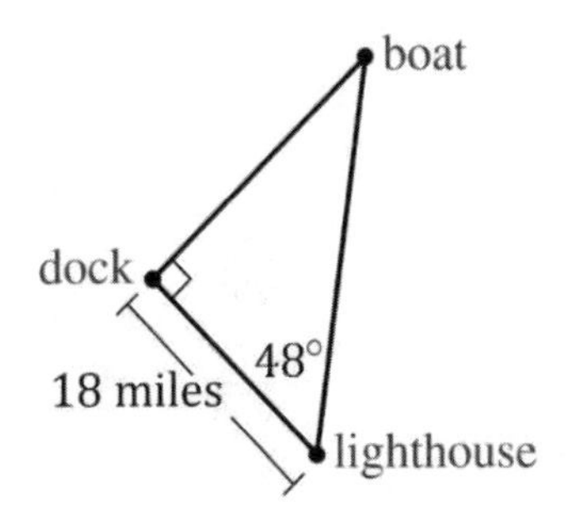

11. The radius of the base of the right circular cone shown below is 9 inches, and the height of the cone is 11 inches. Solving which of the following equations gives the measure, θ, of the angle formed by a slant height of the cone and a radius?

G4 11

a. $\tan\theta = \frac{9}{11}$

b. $\tan\theta = \frac{11}{9}$

c. $\sin\theta = \frac{9}{11}$

d. $\sin\theta = \frac{11}{9}$

e. $\cos\theta = \frac{11}{9}$

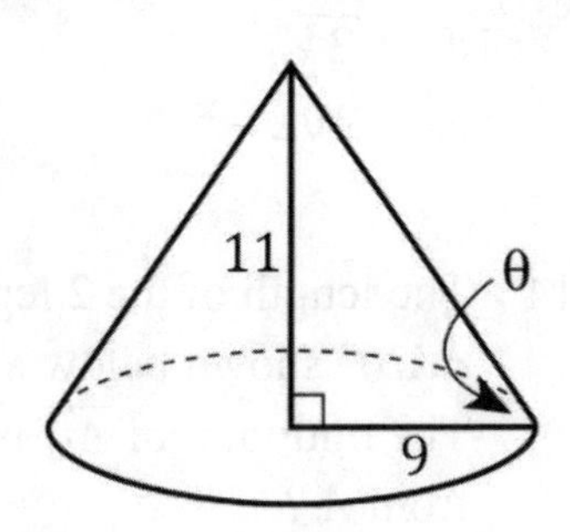

12. For ΔPQR, shown below, which of the following is an expression for x in terms of y ?

G4 12

a. $6 - y$

b. $\sqrt{6 - y^2}$

c. $\sqrt{12 - y^2}$

d. $\sqrt{36 - y^2}$

e. $\sqrt{36 + y^2}$

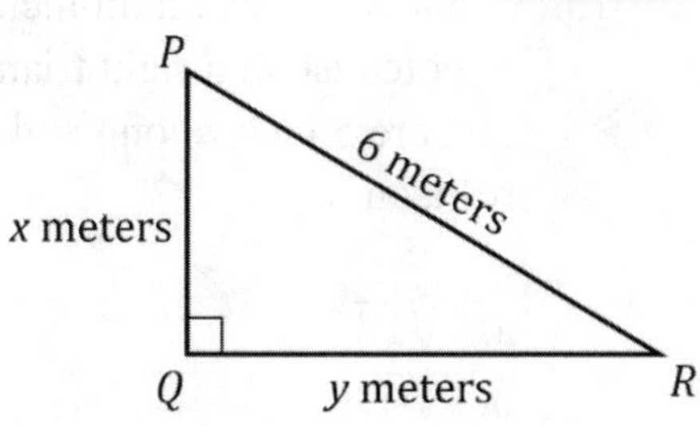

13. In the isosceles right triangle below, $AB = 12$ feet. What is the length, in feet, of $\overline{BC}$?

G4 14

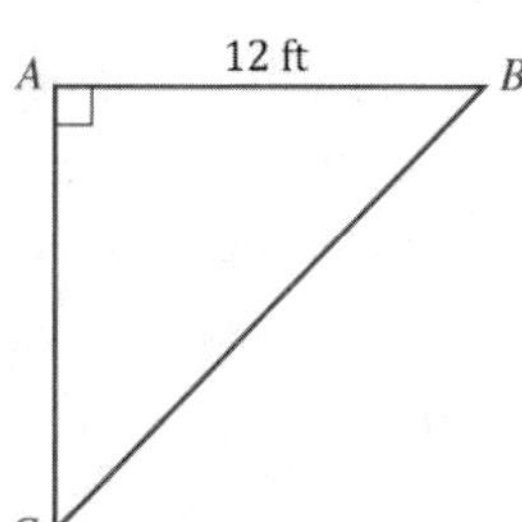

a. 6
b. 12
c. 24
d. $\sqrt{24}$
e. $12\sqrt{2}$

14. The length of the 2 legs of right triangle ΔABC shown below are given in feet. The midpoint of $\overline{AB}$ is how many feet from A ?

G4 13

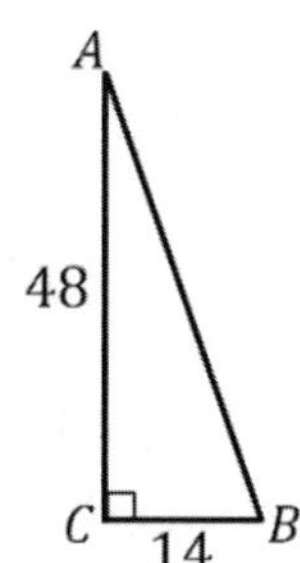

a. 21
b. 25
c. 31
d. 37.5
e. 50

15. What is the length, in inches, of the hypotenuse of a right triangle with legs that are 5 inches long and 8 inches long, respectively?

G4 15

a. $\sqrt{13}$
b. $\sqrt{89}$
c. 13
d. 20
e. 40

FUNCTIONS 1

1. Given $f(x) = \frac{1}{2x}$ and $g(x) = 2x + \frac{1}{x}$, what is $f\left(g\left(\frac{1}{4}\right)\right)$?

   a. $\frac{1}{9}$
   b. $\frac{2}{9}$
   c. $\frac{9}{4}$
   d. $\frac{9}{2}$
   e. 9

2. A function of P is defined as follows:

   For $x > 0,\ P(x) = x^4 + x^3 + 4x + 4$

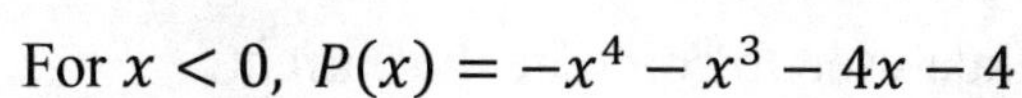

   For $x < 0,\ P(x) = -x^4 - x^3 - 4x - 4$

   What is the value of $P(-2)$?

   a. -20
   b. -4
   c. 0
   d. 4
   e. 20

3. If $f(x) = (4x + 3)^3$, then $f(-1) =$?

F1 04

   a. -37
   b. -27
   c. -3
   d. -1
   e. 1

4. Consider the functions $f(x) = \sqrt{x}$ and $g(x) = 9x - k$. In the standard (x, y) coordinate plane, $y = f\big(g(x)\big)$ passes through $(7, 8)$. What is the value of k ?

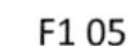

   a. -1
   b. 1
   c. 59
   d. $8 - 9\sqrt{7}$
   e. $7 + 9\sqrt{8}$

5. A formula used to compute the current value of a savings account is $A = P(1 + r)^n$, where A is the current value; P is the amount deposited; r is the rate of interest for 1 compounding period, expressed as a decimal; and n is the number of compounding periods. Which of the following is closest to the value of a savings account after 8 years if \$15,000 is deposited at 3.5% annual interest compounded yearly?

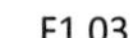

a. \$15,350
b. \$19,752
c. \$48,500
d. \$124,200
e. \$165,486

6. Which of the following describes a true relationship between the functions $f(x) = (x - 4)^2 + 3$ and $g(x) = \frac{1}{3}x + 1$ graphed below in the standard (x, y) coordinate plane?

a. $f(x) = g(x)$ for exactly 2 values of x
b. $f(x) = g(x)$ for exactly 1 value of x
c. $f(x) < g(x)$ for all x
d. $f(x) > g(x)$ for all x
e. $f(x)$ is the inverse of $g(x)$

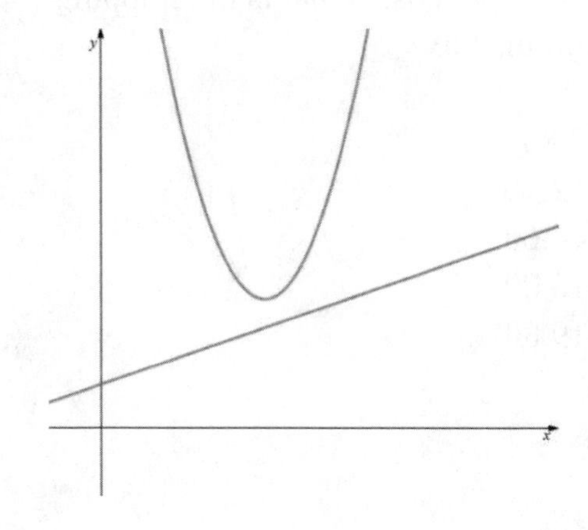

FUNCTIONS 2

1. The shipping rate for customers of Quick Send consists of a fee per box and an additional fee based on the weight, in pounds, of the package. The table below give the fee and the price per pound for customers shipping boxes of various weights.

F2 04

Weight of box (pounds)	Fee	Price per pound
Less than 5	$4.00	$1.00
5-15	$8.00	$0.55
More than 15	$15.00	$0.40

Zafir wants Quick Send to ship 1 box that weighs 12 pounds. What is the shipping rate for this box?

a. $7.80
b. $12.40
c. $14.60
d. $16.00
e. $19.80

2. Darius owns 2 camping stores (X and Y). He stocks 3 brands of tents (A, B, and C) in each store. The matrices below show the numbers of each type of tent in each store and the cost for each type of tent. The value of Darius's tent inventory is computed using the costs listed. What is the total value of the tent inventory for Darius's 2 stores?

F2 03

$$\begin{array}{c} \\ X \\ Y \end{array}\begin{array}{c} \begin{array}{ccc} A & B & C \end{array} \\ \begin{bmatrix} 15 & 30 & 18 \\ 20 & 16 & 35 \end{bmatrix} \end{array} \qquad \begin{array}{c} \\ A \\ B \\ C \end{array}\begin{array}{c} \text{Cost} \\ \begin{bmatrix} \$100 \\ \$250 \\ \$400 \end{bmatrix} \end{array}$$

a. $26,200
b. $30,800
c. $32,400
d. $34,760
e. $36,200

3. Students analyzing a robotic toy observed it move at a constant rate along a straight track. The table below gives the distance, *d* inches, the toy was from a reference point at 1-second intervals from 0 to 5 seconds.

F2 01

t	0	1	2	3	4	5
d	5	11	17	23	29	35

Which of the following equations represents this relationship between d and t ?

a. $d = t + 5$
b. $d = 6t + 3$
c. $d = 6t + 5$
d. $d = 5t + 6$
e. $d = 16t$

4. There are 250 athletes registered for a triathlon, and the competitors are divided into age categories, as shown in the table below.

F2 05

Age Category:	Under 18	18 – 27	28 – 40	Over 40
Number of Competitors:	42	80	95	33

The prize committee has 50 prizes to award and wants the prizes to be awarded in proportion to the number of competitors registered in each category. How many prizes should be designated for the 18 – 27 age category?

a. 8
b. 16
c. 24
d. 32
e. 34

5. Arlo plays basketball on a competitive league in his community. The table below gives Arlo's scoring statistics for last season. How many points did Arlo score playing basketball last season?

F2 02

Type of shot	Number attempted	Percent successful
1-point free throw	60	65%
2-point basket shot	45	80%
3-point basket shot	20	15%

a. 47
b. 60
c. 78
d. 120
e. 210

F2 07

6. The frequency chart below shows the cumulative number of Ms. Maliha's students in her math class whose test scores fell within certain score ranges. All test scores are whole numbers.

Score range	Cumulative number of students
65 – 70	15
65 – 80	17
65 – 90	24
65 – 100	27

How many students have a test score in the interval 81-90 ?

a. 2
b. 3
c. 7
d. 9
e. 10

7. Tip-Top Taxi has an initial fare starting at $1.00, and Checkers Cabs has an initial fare starting at $4.00. Each fare increases a whole number of dollars at each whole number of miles traveled. The graphs below show the passenger fares, in dollars, for both cab companies for trips up to 5 miles. When the fares of the 2 cab companies are compared, what is the cheaper fare for a 4 mile trip?

F2 06

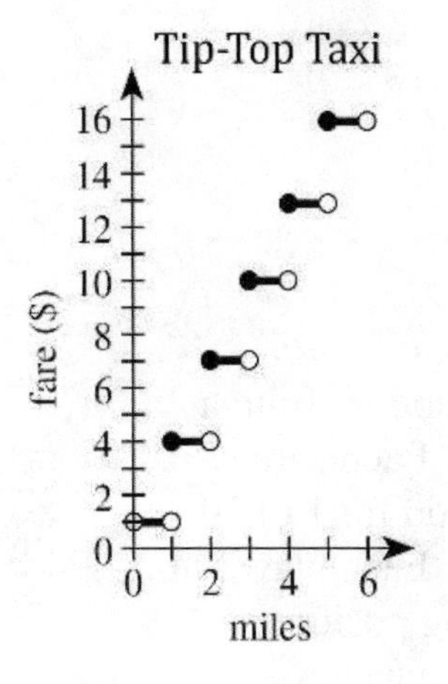

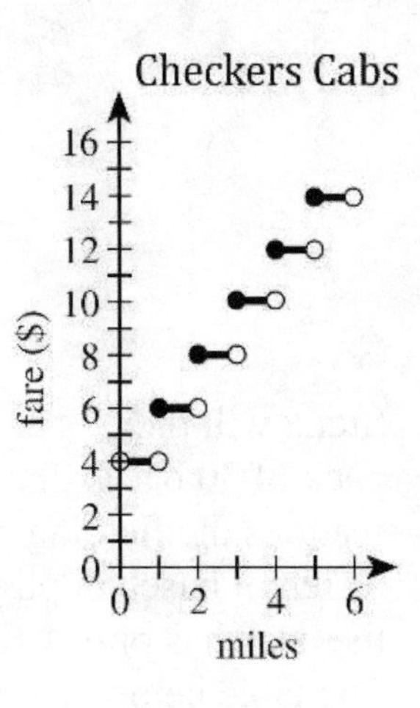

a. $10
b. $11
c. $12
d. $13
e. $14

PROBABILITY

1. Five balls, numbered 1, 2, 3 4, and 5, are placed in a bin. Three balls are drawn at random without replacement. What is the probability that the sum of the numbers on the balls drawn is 10?

P 05

   a. $\frac{1}{5}$
   b. $\frac{2}{5}$
   c. $\frac{1}{4}$
   d. $\frac{2}{9}$
   e. $\frac{2}{7}$

2. Stella will pick 1 card at random from a pack of 30 baseball cards. Each card features the fielding position for 1 of 30 different baseball players. Each player in the pack has only 1 fielding position. The table below lists the frequency of fielding positions in the pack. What is the probability that the card Stella picks will feature a catcher or a pitcher?

P 07

Fielding Position	Frequency
Catcher	8
Infielder	7
Pitcher	10
Outfielder	5

a. 40%
b. 5%
c. 60%
d. 75%
e. 57%

3. The 30-member yearbook club is meeting to choose a student representative. The members decide that the representative, who will be chosen at random, CANNOT be any of the 4 editors of the club. What is the probability that Gina, who is a member of the club but NOT an editor, will be chosen?

P 01

a. 0
b. $\frac{2}{13}$
c. $\frac{1}{30}$
d. $\frac{1}{4}$
e. $\frac{1}{26}$

4. Mr. and Mrs. Smith plan to roof their cabin on 2 consecutive days. The weather forecast is predicting 30% chance of rain each day. Assuming that the chance of rain is independent of the day, what is the probability that it will rain both days?

P 02

a. 0.03
b. 0.06
c. 0.09
d. 0.3
e. 0.6

P 06

5. At the school carnival, Daniel will play a game in which he will draw 3 cards at random from a deck of playing cards. He will be awarded 5 points for each card that is red. Let the variable x represent the total number of points awarded when drawing 3 cards. What is the expected value of x ?

   a. 3
   b. $\frac{5}{2}$
   c. $\frac{15}{2}$
   d. 10
   e. 15

P 03

6. An integer from 1 through 100, inclusive, is to be chosen at random. What is the probability that the number chosen will have 7 as at least 1 digit?

   a. $\frac{7}{100}$
   b. $\frac{10}{100}$
   c. $\frac{18}{100}$
   d. $\frac{19}{100}$
   e. $\frac{19}{99}$

7. In a bag of 500 jelly beans, 20% of the jelly beans are black in color. If you randomly pick a jelly bean from the bag, what is the probability that the jelly bean picked is NOT one of the black jelly beans?

P 04

a. $\frac{1}{2}$

b. $\frac{1}{5}$

c. $\frac{4}{5}$

d. $\frac{1}{20}$

e. $\frac{19}{20}$

STATISTICS

1. What is the difference between the mean and the median of the set $\{6, 12, 14, 24\}$?

S 02

   a. 0
   b. 1
   c. 2
   d. 13
   e. 18

2. What is the median of the following 9 scores? 32, 52, 17, 41, 17, 21, 23, 62, 5

S 04

   a. 17
   b. 22
   c. 23
   d. 27.5
   e. 30

3. Tann is in a bowling competition and has the highest average after 4 games, with scores of 212, 260, 230 and 254. In order to maintain this exact average, what *must* be Tann's score for his 5th game?

S 01

   a. 200
   b. 212
   c. 230
   d. 239
   e. 242

4. The list of numbers 63, 53, 45, x, y, 28 has a median of 40. The mode of the list of numbers is 28. To the nearest whole number, what is the mean of the list?

S 05

   a. 28
   b. 35
   c. 40
   d. 42
   e. 45

5. The average of a list of 4 numbers is 70. A new list of 4 numbers has the same first 3 numbers as the original list, but the fourth number in the original list is 60, and the fourth number in a new list is 80. What is the average of this new list of numbers?

S 03

   a. 70
   b. 71.5
   c. 74
   d. 75
   e. 75.5

6. To increase the mean of 6 numbers by 4, by how much would the sum of the 6 numbers have to increase?

S 06

   a. $\frac{2}{3}$
   b. 1
   c. 12
   d. 24
   e. 36

7. In a small café, customers choose their lunch from 5 sandwiches, 2 soups, 3 salads and 4 drinks. How many different lunches are possible for a customer who chooses exactly 1 sandwich, 1 soup, 1 salad and 1 drink?

S 08

a. 2
b. 4
c. 14
d. 60
e. 120

8. The mean of 5 integers is 29. The median of these 5 integers is 41. Three of the integers are 0, 8 and 41. Which of the following could be one of the other integers?

S 07

a. 16
b. 38
c. 55
d. 58
e. 76

9. During a soccer practice, 8 players position themselves to form an octagon, as seen in the figure below. The players are asked to choose one partner to pass a ball with. How many different partner combinations are possible?

S 09

a. 56
b. 36
c. 28
d. 16
e. 8

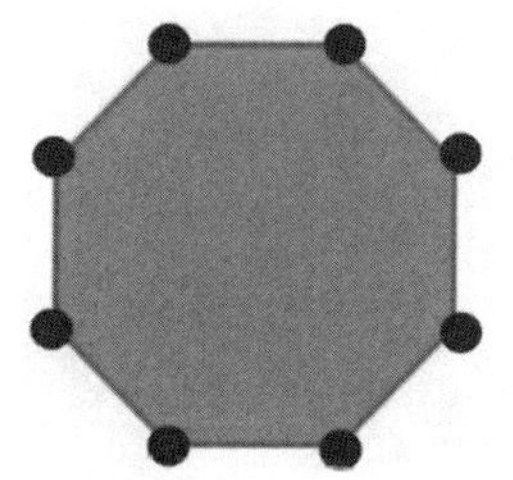

10. The table shows the number of properties a real estate company sold each month last year. What is the median of the data in the table?

S 10

a. 6
b. 9
c. 11
d. 13.5
e. 15.5

Month	Number of homes sold
January	9
February	8
March	11
April	15
May	16
June	19
July	21
August	20
September	15
October	12
November	11
December	6

ACT PRACTICE EXAM 1

1. A wallet containing 6 five-dollar bills, 8 ten-dollar bills, and 10 twenty-dollar bills is found and returned to its owner. The wallet's owner will reward the finder with 1 bill drawn randomly from the wallet. What is the probability that the bill drawn will be a ten dollar bill?

PE 1 01

   a. $\frac{1}{2}$
   b. $\frac{1}{4}$
   c. $\frac{1}{3}$
   d. $\frac{2}{3}$
   e. $\frac{2}{5}$

2. A bag contains 5 blue marbles, 8 black marbles, and 7 orange marbles. How many additional blue marbles must be added to the 20 marbles already in the bag so that the probability of randomly drawing a blue marble is $\frac{3}{4}$?

PE 1 02

   a. 10
   b. 16
   c. 24
   d. 32
   e. 40

3. What is the maximum number of distinct diagonals that can be drawn in the octagon shown below?

PE 1 03

a. 4
b. 8
c. 12
d. 20
e. 28

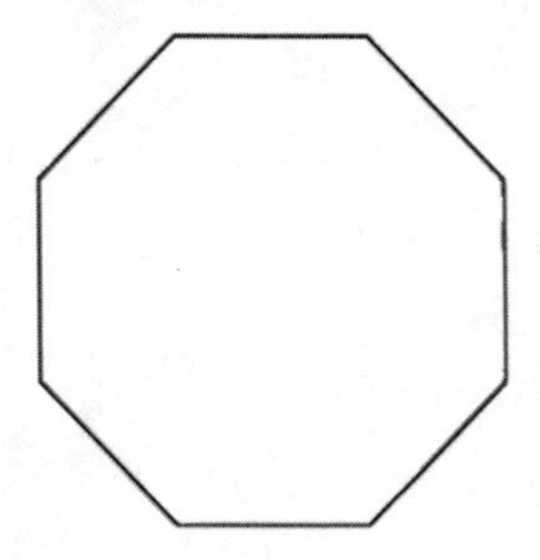

4. A tour group will go from a certain bus stop to a town square by bus on 1 of 5 roads, from the town square to a river by riding bicycles on 1 of 2 bicycle paths, and then from the river to a park by hiking on 1 of 7 trails. How many routes are possible for the tour group to go from the bus stop to the town square to the river to the park?

PE 1 04

a. 7
b. 14
c. 28
d. 56
e. 70

5. Sissy has taken 6 tests in her biology class this semester, and she has an average score of exactly 65 points. How many points does she need to earn on the 7th test to bring her average score up to exactly 70 points?

PE 1 05

a. 100
b. 94
c. 82
d. 70
e. 67.5

6. Tony recently vacationed in Australia. While there, he visited the Sydney Opera House. Afterward, he took a 2.7 kilometer cab ride from the Sydney Opera House to the Sydney Harbor Bridge. Tony's cab ride lasted 12 minutes. Which of the following values is closest to the average speed, in miles per hour, of the cab? (Note: 1 mile = 1.6 kilometers)

PE 1 06

a. 6
b. 8
c. 14
d. 18
e. 22

7. On Tony's vacation in Australia, while on a tour at the Sydney Harbor Bridge, a tour guide named Gaz informed him that 5.9 million rivets were used to build the bridge. When written in scientific notation, the number of rivets used to build the Sydney Harbor Bridge is equal to which of the following expressions?

PE 1 07

a. 5.9×10^6
b. 5.9×10^7
c. 5.9×10^8
d. 59×10^6
e. 59×10^7

8. Tanya, Liza, and Natalie shared a large pizza. Tanya ate $\frac{1}{3}$ of the pizza, Liza ate $\frac{1}{4}$ of the pizza, and Natalie ate the rest. What is the ratio of Tanya's share to Liza's share to Natalie's share?

PE 1 08

a. 1:2:3
b. 2:1:3
c. 3:4:5
d. 4:3:5
e. 5:4:3

PE 1 09

9. As a salesperson at an electronics store, your commission is directly proportional to the dollar amount of sales you make. If your sales are $1200, your commission is $165. How much commission would you earn if you had $2000 in sales?

   a. $295
   b. $275
   c. $255
   d. $250
   e. $180

PE 1 10

10. What is the least common denominator for adding the fractions $\frac{5}{16}, \frac{11}{12},$ and $\frac{7}{15}$?

   a. 180
   b. 240
   c. 480
   d. 960
   e. 2880

11. What is the value of the following expression, $||-9+6|-|2-10||$?

PE 1 11

a. –11
b. –5
c. 0
d. 5
e. 11

12. Which of the following is an irrational number?

PE 1 12

a. $\sqrt{1}$
b. $\sqrt{49}$
c. $\sqrt{121}$
d. $\sqrt{\frac{5}{25}}$
e. $\sqrt{\frac{81}{16}}$

13. A hotel manager is considering raising the room rates by 32%. What will the new rate be for a room that currently costs $60.00 ?

PE 1 13

a. $60.32
b. $63.20
c. $79.00
d. $79.20
e. $92.00

14. The *determinant* of a matrix $\begin{bmatrix} a & b \\ c & d \end{bmatrix}$ equals $ad - cb$. What must be the value of x for the matrix $\begin{bmatrix} x & x \\ 12 & x \end{bmatrix}$ to have a determinant of -36 ?

PE 1 14

a. –6

b. –3

c. $-\frac{12}{7}$

d. $\frac{12}{5}$

e. 6

15. The 6 consecutive integers below add up to 375.

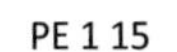

PE 1 15

$x - 4$

$x - 3$

$x - 2$

$x - 1$

x

$x + 1$

What is the value of x ?

a. 62
b. 63
c. 64
d. 65
e. 66

16. If $\frac{5\sqrt{2}}{x\sqrt{2}} = \frac{5\sqrt{2}}{2}$ is true, then x = ?

PE 1 16

a. 1
b. $\sqrt{2}$
c. 2
d. 4
e. 10

17. The sum of the real numbers x and y is 15. Their difference is 7. What is the value of xy?

PE 1 17

a. 4
b. 7
c. 11
d. 44
e. 105

18. If $a = 1, b = -8$, and $c = 6$, what does $(a - b + c)(b + c)$ equal?

PE 1 18

a. -30
b. -2
c. 2
d. 13
e. 30

19. Given that y varies directly as the cube of x, if $y = 81$ when $x = 3$, what is y when $x = 2$?

PE 1 19

a. 16
b. 24
c. 27
d. 54
e. 78

20. A cleaning company charges $32 for each hour they work on cleaning a house, plus a flat $20 booking fee. How many hours of work are included in a $212 bill for a consultation?

PE 1 20

a. 4
b. $4\frac{1}{13}$
c. 6
d. $6\frac{5}{8}$
e. 9

21. This month, Andre sold 135 doll houses in 2 sizes. The large doll houses sold for $11 each, and the small doll houses sold for $4 each. The amount of money he received from the sales of the large doll houses was equal to the amount of money he received from the sales of small doll houses. How many large doll houses did Andre sell this month?

PE 1 21

a. 9
b. 15
c. 36
d. 44
e. 99

22. The text message component of each of Ruby's monthly phone bills consists of $12.00 for the first 250 text messages sent that month, plus $0.15 for each additional text message sent that month. On Ruby's most recent phone bill she was charged a total of $24.75 for text messages. For how many text messages in total was Ruby charged on the bill?

PE 1 22

a. 165
b. 315
c. 335
d. 350
e. 415

23. Mr. Lee plans to drive 880 miles to his cabin, driving an average of 44 miles per hour. How many miles per hour faster must he average while driving, to reduce his total driving time by 4 hours?

PE 1 23

a. 4
b. 11
c. 14
d. 16
e. 20

24. For all x in the domain of the function $\frac{x+3}{x^3-9x}$, this function is equivalent to:

PE 1 24

a. $\frac{1}{x^2} - \frac{1}{x^3}$

b. $\frac{1}{x^3} - \frac{1}{9x}$

c. $\frac{1}{x^2-3}$

d. $\frac{1}{x^2-3x}$

e. $\frac{1}{x^3}$

25. Given $p = 6q^4 - 30$, which of the following is an expression for q in terms of p ?

PE 1 25

a. $\left(\frac{p}{6} - 5\right)^{\frac{1}{4}}$

b. $\left(\frac{p}{6} + 5\right)^{\frac{1}{4}}$

c. $\frac{1}{6}(p + 5)^{\frac{1}{4}}$

d. $p^4 - 5$

e. $6p^4 - 30$

26. The product $(5x^3y^7)(2x^4y)$ is equivalent to:

PE 1 26

a. $7x^7y^8$
b. $10x^7y^7$
c. $10x^7y^8$
d. $7x^{12}y^7$
e. $10x^{12}y^7$

27. A cat eats 5 cups of food in 4 days. At this rate, how many cups of food does the cat eat in $4 + d$ days?

PE 1 27

a. $\frac{5}{4} + d$

b. $\frac{5}{4} + \frac{d}{4}$

c. $\frac{5}{4} + \frac{5}{4d}$

d. $5 + \frac{d}{4}$

e. $5 + \frac{5d}{4}$

28. What polynomial must be added to $4x^2 - 5x + 8$ so that the sum is $5x^2 + 4x$?

PE 1 28

a. $9x^2 - x + 8$
b. $5x^2 + 9x + 8$
c. $5x^2 + 9x - 8$
d. $x^2 + 9x - 8$
e. $x^2 + x + 8$

29. Examine the following system of equations:

PE 1 29

$y = x^2$ $\qquad\qquad$ $px - r = qy$

where $p, q,$ and r are integers. For which of the following will there be no (x, y) solutions for the system?

a. $p^2 - 4qr < 0$
b. $q^2 - 4pr < 0$
c. $p^2 + 4qr > 0$
d. $q^2 - 4pr > 0$
e. $q^2 + 4pr > 0$

30. For what value of r would be the following system of equations have an infinite number of solutions?

PE 1 30

$$3p - 2q = 6$$

$$12p - 8q = 2r$$

a. 4
b. 8
c. 12
d. 24
e. 30

31. If $x < |y|$, which of the following is the solution statement for x when $y = -8$?

PE 1 31

a. x is any real number.
b. $x > 8$
c. $x < 8$
d. $-8 < x < 8$
e. $x > 8$ or $x < 8$

32. Which of the following is the equation of a circle centered at $(0, 0)$ that passes through the point $(25, 0)$?

PE 1 32

a. $(x + y)^2 = 25$
b. $(x + y)^2 = 25^2$
c. $x^2 + y^2 = 5$
d. $x^2 + y^2 = 25$
e. $x^2 + y^2 = 25^2$

33. What is the slope of the line through $(-3, 4)$ and $(3, -6)$ in the standard (x, y) coordinate plane?

PE 1 33

a. $\frac{5}{3}$
b. 1
c. $-\frac{1}{3}$
d. $-\frac{5}{3}$
e. -2

34. In the standard (x, y) coordinate plane, D' is the image resulting from the reflection of the point $D(6, -5)$ across the y-axis. What are the coordinates of D' ?

PE 1 34

a. $(6, 5)$
b. $(-6, -5)$
c. $(-6, 5)$
d. $(5, 6)$
e. $(5, -6)$

35. The graph of $y = |x + 4|$ is in the standard coordinate plane. Which of the following transformations, when applied to the graph of $y = |x|$, results in the graph of $y = |x + 4|$?

PE 1 35

a. Translation to the right 4 coordinate units
b. Translation to the left 4 coordinate units
c. Translation up 4 coordinate units
d. Translations down 4 coordinate units
e. Reflection across the line $x = 4$

36. The coordinates of the endpoints of $\overline{AB}$, in the standard (x, y) coordinate plane, are $(-6, -2)$ and $(14, 4)$. What is the x-coordinate of the midpoint of $\overline{AB}$?

PE 1 36

a. 1
b. 2
c. 4
d. 8
e. 10

37. If point A has a nonzero x-coordinate and a nonzero y-coordinate and the coordinates have the same signs, then point A *must* be located in which of the 4 quadrants labeled below?

PE 1 37

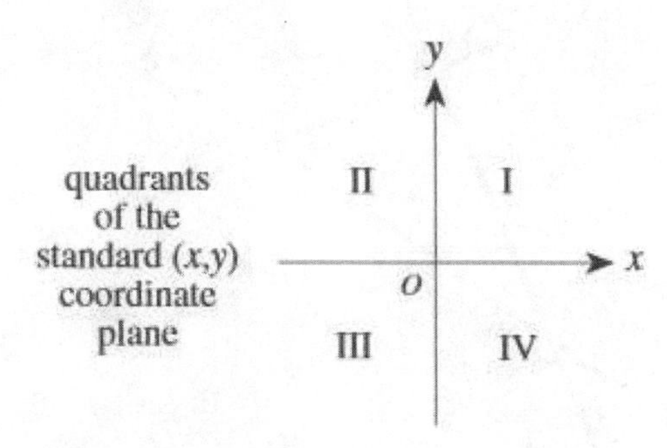

a. I only
b. III only
c. I or III only
d. I or IV only
e. II or IV only

38. When graphed in the standard (x, y) coordinate plane, the lines $y = -2$ and $y = x + 3$ intersect at what point?

PE 1 38

a. $(1, -2)$
b. $(-2, -5)$
c. $(-2, -2)$
d. $(-2, 3)$
e. $(-5, -2)$

39. In the figure below, $\overleftrightarrow{EB}$ intersects $\overleftrightarrow{DF}$ at A and is perpendicular to $\overleftrightarrow{BG}$. Line $\overleftrightarrow{BG}$ intersects at C. Given that the measure of $\angle FCG$ is 46°, what is the measure of $\angle EAC$?

PE 1 39

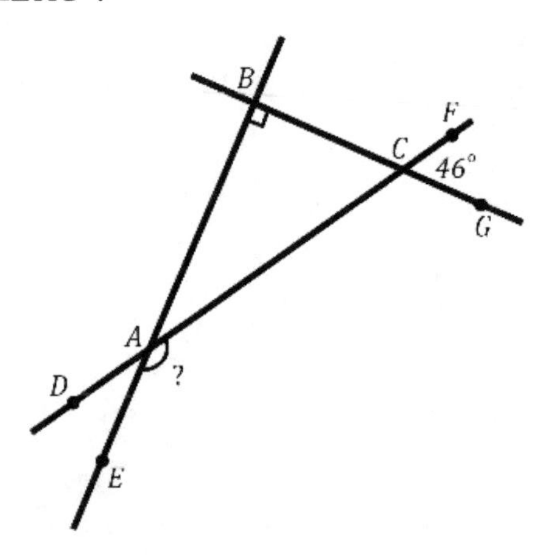

a. 44°
b. 92°
c. 132°
d. 136°
e. 165°

40. In any parallelogram $PQRS$, it is always true that the measures of $\angle PQR$ and $\angle RSP$:

PE 1 40

a. are supplementary
b. are complementary
c. are congruent
d. are each 90°
e. are each less than 90°

41. In a plane, the distinct lines $\overleftrightarrow{PQ}$ and $\overleftrightarrow{RS}$ intersect at P, where P is between R and S. The measure of $\angle QPR$ is 42°. What is the measure of $\angle QPS$?

PE 1 41

a. 42°
b. 48°
c. 84°
d. 138°
e. 142°

42. Bass Treble Custom Sound is designing a packing box for its new line of Tenaciously Crisp speaker systems. The box is a rectangular prism of length 65 centimeters, width 40 centimeters, and volume 208,000 cubic centimeters. What is the height, in centimeters, of the box?

PE 1 42

a. 105
b. 80
c. 75
d. 60
e. 45

Use the following information to answer the next 3 questions.

The figure below shows the design of a circular black-and-white dartboard at a local hobby shop. The diameter of the dartboard is 16 inches.

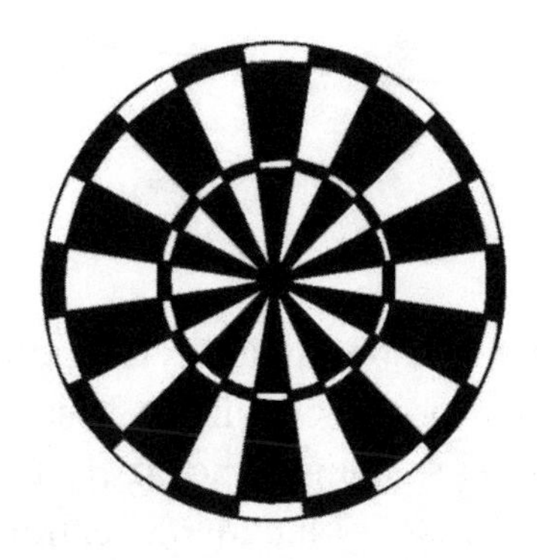

43. The design of the dartboard has how many lines of symmetry?

PE 1 43

a. 2
b. 4
c. 10
d. 20
e. Infinitely many

44. What is the area of the dartboard, to the nearest 0.1 square inch?

PE 1 44

a. 50.2
b. 78.9
c. 100.5
d. 201.0
e. 803.8

45. Doug wants to order a larger dartboard for his children to use. Its design will be identical to that of the dartboard from the hobby shop. However, the new dartboard will be 75% longer. The new dartboard will be how many inches in diameter?

a. 12
b. 20
c. 22
d. 28
e. 32

46. A right circular cylinder is shown in the figure below, with dimensions given in centimeters. What is the total surface area of this cylinder, in square centimeters?

(Note: The total surface area of a cylinder is given by $2\pi r^2 + 2\pi rh$ where r is the radius and h is the height.)

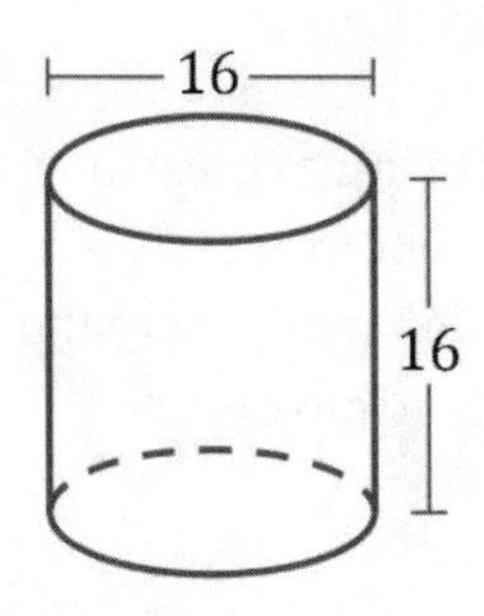

a. 128π
b. 256π
c. 288π
d. 320π
e. 384π

47. In the right triangle ΔPRS below, $\overline{QT}$ is parallel to $\overline{RS}$, and $\overline{QT}$ is perpendicular to $\overline{PS}$ at T. The length of $\overline{PR}$ is 32.5 feet, the length of $\overline{QT}$ is 5 feet, and the length of $\overline{PT}$ is 12 feet. What is the length, in feet, of $\overline{RS}$?

PE 1 47

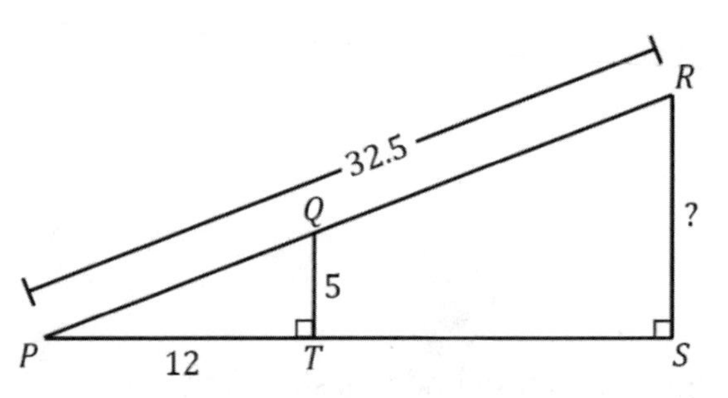

a. 10
b. 12.5
c. 16.25
d. 17
e. 20.5

48. Which of the following expressions represents the number of meters a runner must travel in a 4-lap race where the course is a circle of radius R meters?

PE 1 48

a. $4R$
b. $4\pi R$
c. $4\pi R^2$
d. $8R$
e. $8\pi R$

49. Rishma pounded a stake into the ground. When she attached a leash to both the stake and her dog's collar, the dog could reach 12 feet from the stake in any direction. Using 3.14 for π, what is the approximate area of the lawn, in square feet, the dog could reach from the stake?

PE 1 49

a. 38
b. 75
c. 151
d. 452
e. 504

50. What is the period of the function $f(x) = \csc(6x)$?

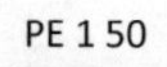

PE 1 50

a. π

b. 2π

c. 6π

d. $\frac{\pi}{6}$

e. $\frac{\pi}{3}$

51. The dimensions of the right triangle shown below are given in inches. What is $\cos\theta$?

PE 1 51

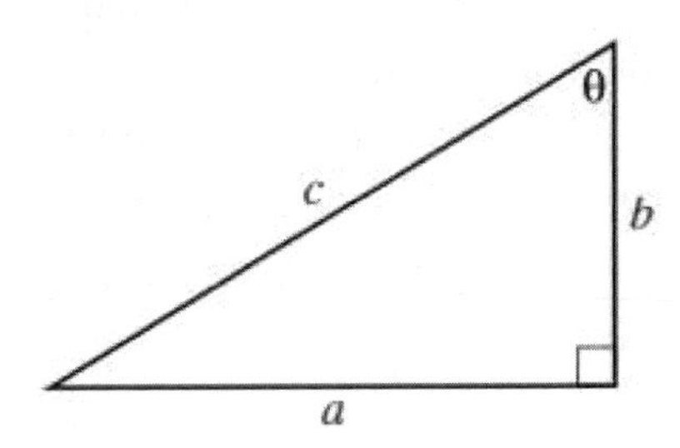

a. $\frac{a}{b}$

b. $\frac{a}{c}$

c. $\frac{b}{c}$

d. $\frac{b}{a}$

e. $\frac{c}{b}$

52. What is the perimeter, in inches, of an isosceles right triangle whose hypotenuse is $3\sqrt{8}$ inches long?

PE 1 52

a. 3
b. $3 + 3\sqrt{8}$
c. $3 + 6\sqrt{8}$
d. 6
e. $12 + 3\sqrt{8}$

53. In ΔABC, shown below, the measure of $\angle B$ is 49°, the measure of $\angle C$ is 42°, and $\overline{AB}$ is 16 feet long. Which of the following is an expression for the length, in feet, of $\overline{BC}$?

PE 1 53

(Note: The law of sines states that, for any triangle, the ratios of the sines of the interior angles to the lengths of the sides opposite those angles are equal.)

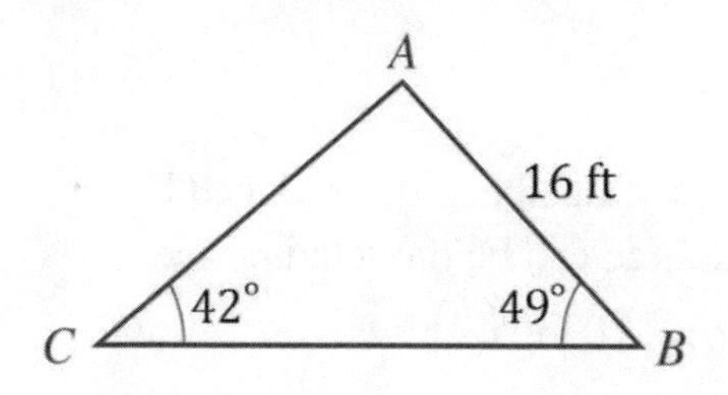

a. $\frac{16 \sin 89°}{\sin 49°}$

b. $\frac{16 \sin 89°}{\sin 42°}$

c. $\frac{16 \sin 91°}{\sin 49°}$

d. $\frac{16 \sin 49°}{\sin 89°}$

e. $\frac{16 \sin 42°}{\sin 91°}$

54. Given $f(x) = 2x^2 - 5x$ and $g(x) = x - 3$, what is $f(g(x))$?

PE 1 54

a. $2x^2 - 5x - 3$
b. $2x^2 - 17x + 3$
c. $2x^2 - 17x + 33$
d. $2x^3 - 11x^2 + 15x$
e. $2x^4 - 5x^3 + 15x^2$

55. A function, f, is defined by $f(x, y) = 5x^2 - 3y$. What is the value of $f(3, 5)$?

PE 1 55

a. 15
b. 30
c. 75
d. 116
e. 210

56. The number of decibels, d, produced by an audio source can be modeled by the equation $d = 10\log\left(\frac{I}{K}\right)$, where I is the sound intensity of the audio source and K is a constant. How many decibels are produced by an audio source whose sound intensity is 10,000 times the value of K ?

PE 1 56

a. 5
b. 40
c. 50
d. 1000
e. 100,000

57. Christina is training for a marathon. She plans to run 3 miles in her first training session, then continually increase the distance by a $\frac{1}{2}$ mile for each training session afterwards. Christina created the table below to monitor her progress.

PE 1 57

Training session (t)	1	2	3	4	5	6
Distance (d)	3	3.5	4	4.5	5	5.5

Which of the following equations represents this data?

a. $d = 0.5t + 1.5$
b. $d = 0.5t + 2.5$
c. $d = t + 2$
d. $d = 3t - 2.5$
e. $d = 3t$

Use the following information to answer the next 3 questions.

Skye makes and sells wooden picture frames in 2 sizes: small and large. It takes him 1 hour to make a small frame and 2 hours to make a large frame. The shaded triangular region shown below is the graph of a system of inequalities representing weekly constraints Skye has in making the frames. For making and selling s small frames and l large frames, Skye makes a profit of $35s + 80l$ dollars. Skye sells all the frames he makes.

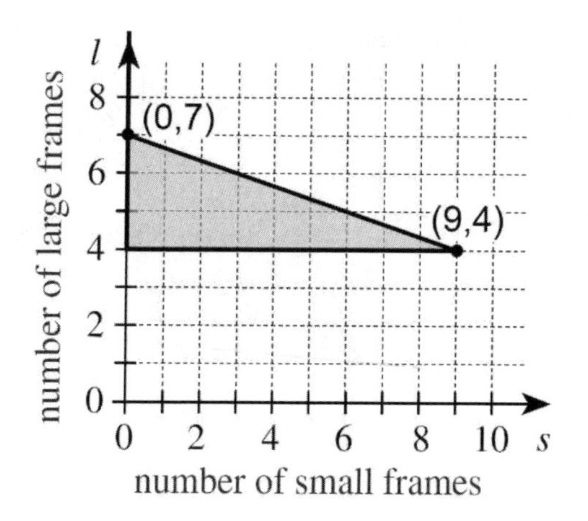

PE 1 58

58. The weekly constraint represented by the horizontal line segment containing (9, 4) means that each week Skye makes a minimum of:

    a. 4 large frames.
    b. 9 large frames.
    c. 4 small frames.
    d. 9 small frames.
    e. 13 large frames.

59. For every hour that Skye spends making frames in the first week of December each year, he donates \$8 from that week's profits to a local charity. This year, Skye made 5 large frames and 6 small frames in that week. Which of the following is closest to the percent of that week's profit Skye donated to the charity?

PE 1 59

a. 11%
b. 14%
c. 16%
d. 20%
e. 21%

60. What is the maximum profit Skye can earn from the picture frames he makes in 1 week?

PE 1 60

a. \$560
b. \$585
c. \$610
d. \$635
e. \$660

ACT PRACTICE EXAM 2

1. As part of a pop quiz, Ricardo is to answer 3 multiple-choice questions. For each question, there are 5 possible answers, only 1 of which is correct. If Ricardo randomly and independently answers each question, what is the probability that he will answer the 3 questions correctly?

PE 2 01

a. $\frac{15}{125}$

b. $\frac{25}{125}$

c. $\frac{4}{125}$

d. $\frac{5}{125}$

e. $\frac{1}{125}$

2. Only tenth-, eleventh-, and twelfth-grade students attend Sir Winston High School. The ratio of tenth graders to the school's total student population is 72:273, and the ratio of eleventh graders to the school's total student population is 14:39. If 1 student is chosen at random from the entire school, which grade is that student most likely to be in?

PE 2 02

a. Tenth
b. Eleventh
c. Twelfth
d. All grades are equally likely.
e. Cannot be determined from the given information

3. What is the sum of the mean and the median of the first 10 prime numbers?

PE 2 03

a. 12
b. 24.9
c. 27.9
d. 129
e. 154.8

4. A wedding organizer is deciding how to arrange 6 guests around a specific dinner table. How many different arrangements are possible?

PE 2 04

a. 21
b. 24
c. 36
d. 120
e. 720

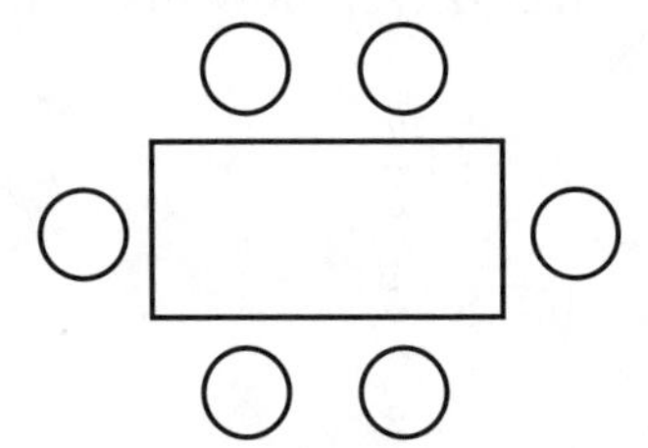

5. The monthly fees for one-bedroom apartments at 5 different apartment buildings are \$560, \$610, \$580, \$640, and \$520, respectively. What is the mean of these monthly fees?

PE 2 05

a. \$550
b. \$580
c. \$582
d. \$590
e. \$620

6. A construction company that builds bridges used a pile driver to drive a post into the ground. The post was driven 24 feet into the ground by the first hit of the piledriver. On each hit after the first hit, the post was driven into the ground an additional distance that was $\frac{3}{4}$ the distance the post was driven in the previous hit. After a total of 4 hits, the post was driven how many feet into the ground?

PE 2 06

a. $31\frac{7}{8}$
b. $55\frac{1}{2}$
c. $65\frac{5}{8}$
d. 72
e. 84

7. $\frac{5}{\sqrt{3}} + \frac{2}{\sqrt{5}} = ?$

PE 2 07

a. $\frac{5\sqrt{5}+2\sqrt{3}}{\sqrt{8}}$

b. $\frac{5\sqrt{5}+2\sqrt{3}}{\sqrt{15}}$

c. $\frac{7}{\sqrt{3}+\sqrt{5}}$

d. $\frac{7}{\sqrt{8}}$

e. $\frac{10}{\sqrt{15}}$

8. In what order should $\frac{7}{5}, \frac{11}{8}, \frac{5}{4}$, and $\frac{4}{3}$ be listed to be arranged by increasing size?

PE 2 08

a. $\frac{4}{3} < \frac{5}{4} < \frac{7}{5} < \frac{11}{8}$

b. $\frac{11}{8} < \frac{7}{5} < \frac{5}{4} < \frac{4}{3}$

c. $\frac{4}{3} < \frac{5}{4} < \frac{11}{8} < \frac{7}{5}$

d. $\frac{5}{4} < \frac{4}{3} < \frac{11}{8} < \frac{7}{5}$

e. $\frac{5}{4} < \frac{4}{3} < \frac{7}{5} < \frac{11}{8}$

9. Which of the following lists all the positive factors of 16?

PE 2 09

a. 1, 16
b. 2, 4
c. 2, 4, 6
d. 16, 32, 48
e. 1, 2, 4, 8, 16

PE 2 10

10. If the length of a rectangle is increased by 30% and the width is decreased by 15%, the area of the resulting rectangle is larger than the area of the original rectangle by what percent?

    a. 4.5%
    b. 10.5%
    c. 15%
    d. 19.5%
    e. 45%

PE 2 11

11. Rachelle left her home at 10:00 a.m. on Friday and traveled 672 miles. When she arrived at her destination it was 2:00 a.m. the next day. Given that her home and her destination are in the same time zone, which of the following is closest to her average speed, in miles per hour, for this trip?

    a. 56
    b. 48
    c. 42
    d. 28
    e. 16

12. The first term is 2 in the geometric sequence 2, -8, 32, -128,… What is the SEVENTH term of the geometric sequence?

PE 2 12

a. -2048
b. -40
c. 512
d. 1028
e. 8192

13. If there are 9×10^8 helium molecules in a volume of 3×10^4 cubic centimeters, what is the average number of helium molecules per cubic centimeter?

PE 2 13

a. 4×10^{-5}
b. 3×10^2
c. 3×10^4
d. 27×10^{12}
e. 27×10^{32}

14. In real numbers, what is the solution of the equation $8^{3x-2} = 16^{x-2}$?

PE 2 14

a. $-\frac{14}{5}$
b. $-\frac{2}{5}$
c. 0
d. $\frac{2}{5}$
e. $\frac{13}{14}$

15. The equation $|4x - 12| - 2 = 6$ has 2 solutions. Those solutions are equal to the solutions to which of the following pairs of equations?

PE 2 15

a. $4x - 14 = 6$
$-4x - 14 = -6$

b. $4x - 12 = 8$
$-4x - 12 = 8$

c. $4x - 12 = 4$
$-(4x - 12) = 4$

d. $4x - 12 = 8$
$-(4x - 12) = 4$

e. $4x - 12 = 8$
$-(4x - 12) = 8$

16. For what positive real value of x, if any, is the determinant of the matrix $\begin{bmatrix} 8 & x \\ x & 2 \end{bmatrix}$ equal to $6x$?

PE 2 16

(Note: The determinant of matrix $\begin{bmatrix} a & b \\ c & d \end{bmatrix}$ equals $ad - bc$.)

a. 2
b. 4
c. 8
d. 16
e. There is no such value of x.

17. Which of the following is equivalent to the inequality $6x - 4 < 9x + 8$?

PE 2 17

a. $x < -4$
b. $x > -4$
c. $x < 2$
d. $x > 5$
e. $x > 4$

18. When $x = \frac{1}{5}$, what is the value of $\frac{15x-2}{x}$?

PE 2 18

a. $\frac{1}{5}$

b. 1

c. $\frac{13}{5}$

d. 5

e. 13

19. A group of cells grow in number as described by the equation $y = 8(2)^t$, where y represents the number of cells after t days. According to this formula, how many cells will be in the group at the end of the first 7 days?

PE 2 19

a. 56
b. 112
c. 392
d. 512
e. 1024

20. Wired Connection, an internet provider, charges each customer \$60 for installation, plus \$75 per month for internet service. Wired Connection's competitor, Quicker Cables, charges each customer \$20 for installation, plus \$80 per month for internet service. A customer who signs up with Wired Connection will pay the same total amount for internet as a customer who signs up with Quicker Cables if each pays for installation and internet service for how many months?

PE 2 20

a. 4
b. 5
c. 8
d. 10
e. 16

21. For 2 consecutive integers, the result of adding the smaller integer and one third of the larger integer is 31. What are the 2 integers?

PE 2 21

a. 15, 16
b. 22, 23
c. 23, 24
d. 24, 25
e. 30, 31

22. Noa sold 8 used video games for \$14.50 each. With the money from these sales, she bought 3 new video games and had \$26.15 left over. What was the average amount Noa paid for each new video game?

PE 2 22

a. \$24.50
b. \$29.95
c. \$38.67
d. \$43.50
e. \$47.38

23. A rectangle has an area of 60 square feet and a perimeter of 32 feet. What is the shortest of the side lengths, in feet, of the rectangle?

PE 2 23

a. 2
b. 4
c. 6
d. 10
e. 15

24. Newton's law of universal gravitation is represented by $F = \frac{Gm_1m_2}{r^2}$ where F is the force, G is the gravitational constant, m, and m_2 are the masses of two objects, and r is the distance between the centers of the objects. Which of the following is an expression for m_1, in terms of F, G, m_2, and r ?

PE 2 24

a. FGm_2r^2

b. $\frac{Gm_2}{Fr^2}$

c. $\frac{r^2}{FGm_2}$

d. $\frac{Fr^2}{Gm_2}$

e. $\frac{FGr^2}{m_2}$

25. The expression $(5x^2 + 3y)(5x^2 - 3y)$ is equivalent to:

PE 2 25

a. $25x^4 - 9y^2$
b. $25x^4 - 6y^2$
c. $25x^4 + 9y^2$
d. $10x^4 - 9y^2$
e. $10x^4 - 6y^2$

26. $(3x + 4y + z) - (5x + 8y - 6z)$ is equivalent to:

PE 2 26

a. $-5x - 12y - 5z$
b. $-5x - 4y + 7z$
c. $-2x + 12y - 5z$
d. $-2x - 4y - 5z$
e. $-2x - 4y + 7z$

27. What is the product of the complex numbers $(-4i + 6)$ and $(4i + 6)$?

PE 2 27

a. 1
b. 20
c. 52
d. $-20 + 40i$
e. $20 + 48i$

28. The daily fee for staying at the Sunny Daze Campground is \$8 per vehicle and \$4 per tent. Last month, daily fees were paid for v vehicles and t tents. Which of the following expressions gives the total amount, in dollars, collected for daily fees last month?

PE 2 28

a. $8v + 4t$
b. $8t + 4v$
c. $4(v + t)$
d. $12(v + t)$
e. $4(v + t) + 8t$

29. If a and b are real numbers such that $a < -1$ and $b > 1$, then which of the following inequalities *must* be true?

PE 2 29

a. $a^2 - 3 < b^2 - 3$

b. $a^{-2} < b^{-2}$

c. $|a| < |4b|$

d. $2a + 6 < 2b + 6$

e. $\frac{a}{b} > 1$

30. Two numbers are *reciprocals* if their product is equal to 1. If x and y are reciprocals and $0 < x < 1$, then y must be:

PE 2 30

a. less than -1.
b. between 0 and -1.
c. equal to 0.
d. between 0 and 1.
e. greater than 1.

31. Which of the following is false for all consecutive integers x and y such that $x > y$?

PE 2 31

a. x is odd
b. y is even
c. $x - y = 1$
d. $x^2 - y^2$ is odd
e. $x^2 + y^2$ is even

32. In the standard (x, y) coordinate plane, what is the slope of the line given by the equation $9x = 5y - 2$?

PE 2 32

a. $-\frac{9}{5}$

b. $\frac{9}{5}$

c. $\frac{5}{9}$S

d. 9

e. 5

33. The sides of a square are 4 cm long. One vertex of the square is at (0, -3) on a square coordinate grid marked in centimeter units. Which of the following points could also be a vertex of the square?

PE 2 33

a. (-3, 0)
b. (-1, -4)
c. (-1, -2)
d. (0, 1)
e. (0, 7)

34. In the standard (x, y) coordinate plane, the center of the circle shown below lies on the x-axis at $x = -3$. If the circle is tangent to the y-axis, which of the following is an equation of the circle?

PE 2 34

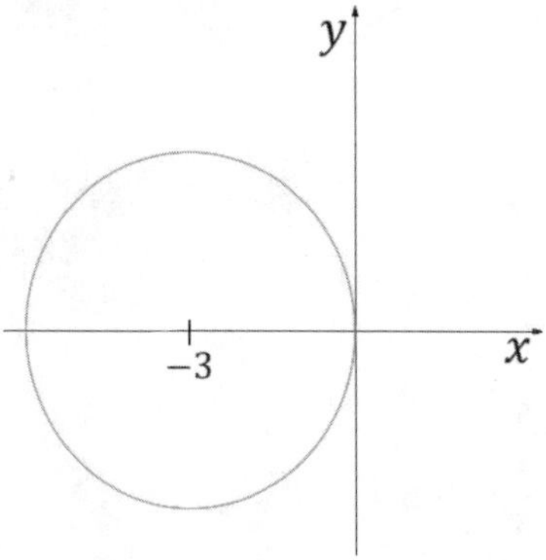

a. $(x-3)^2 + y^2 = 3$
b. $(x+3)^2 + y^2 = 9$
c. $(x+3)^2 - y^2 = 9$
d. $(x+3)^2 + y^2 = 3$
e. $x^2 + (y+3)^2 = 9$

35. In the standard (x, y) coordinate plane, the midpoint of $\overline{AB}$ is at $(3, 6)$, and A is at $(12, 8)$. What is the x-coordinate of B ?

PE 2 35

a. -6
b. -3
c. 0
d. 4
e. 9

36. The graph in the standard (x, y) coordinate plane below is represented by which of the following equations?

PE 2 36

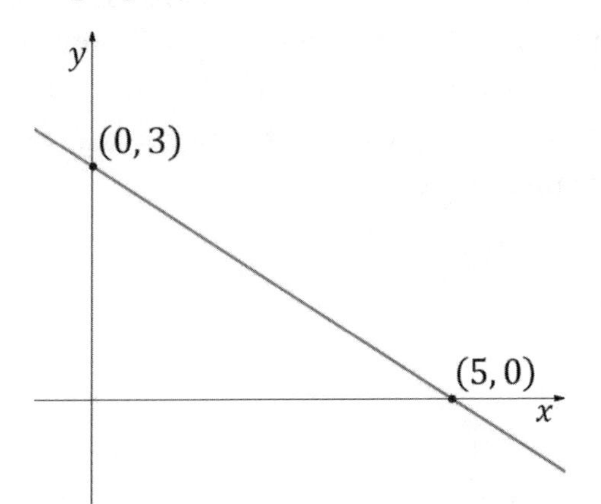

a. $y = -\frac{3}{5}x + 3$

b. $y = -\frac{3}{5}x + 5$

c. $y = -\frac{5}{3}x + 3$

d. $y = -\frac{5}{3}x + 5$

e. $y = \frac{3}{5}x + 3$

37. The graph of $y = 4x^2 - 10$ passes through $(-1, 3a)$ in the standard (x, y) coordinate plane. What is the value of a ?

PE 2 37

a. -9
b. -2
c. 2
d. 3
e. 5

38. A point at $(4, -6)$ in the standard (x, y) coordinate plane is translated left 6 coordinate units and up 4 coordinate units. What are the coordinates of the point after the translation?

PE 2 38

a. $(10, -10)$
b. $(0, 0)$
c. $(-2, -2)$
d. $(-2, -10)$
e. $(-10, -10)$

39. In the figure below, $\overline{AB} \parallel \overline{CD}$, $\overline{AE}$ bisects $\angle BAC$, and $\overline{CE}$ bisects $\angle ACD$. If the measure of $\angle ACD$ is 94°, what is the measure of $\angle AEC$?

PE 2 39

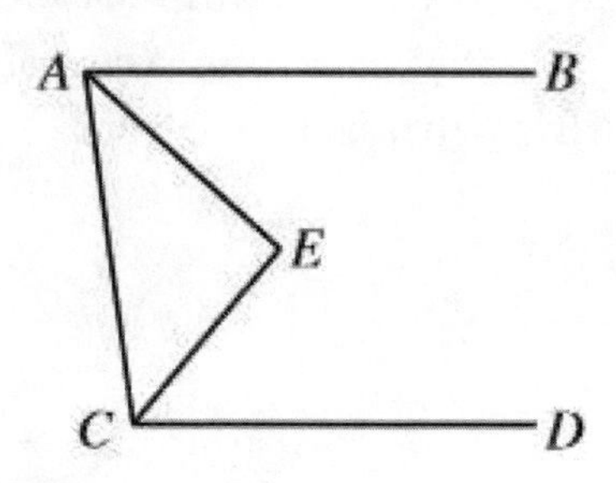

a. 86°
b. 88°
c. 90°
d. 92°
e. Cannot be determined from the given information

40. In the figure below, C is the intersection of $\overline{AD}$ and $\overline{BE}$. If it can be determined, what is the measure of $\angle BAC$?

PE 2 40

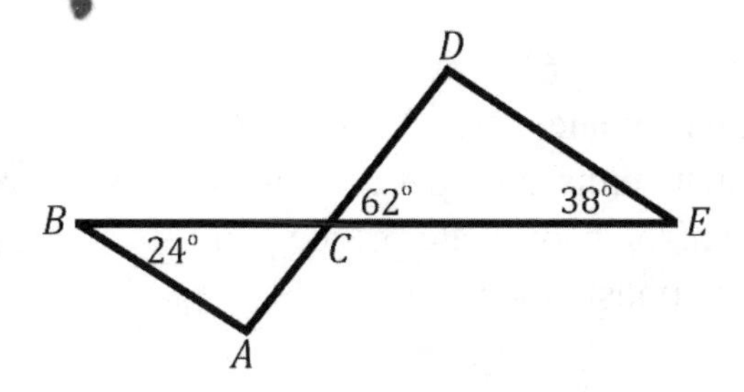

a. 90°
b. 94°
c. 100°
d. 124°
e. Cannot be determined from the given information

41. In the figure shown below, the measure of $\angle BAC$ is $(x - 35)°$ and the measure of $\angle BAD$ is 90°. What is the measure of $\angle CAD$?

PE 2 41

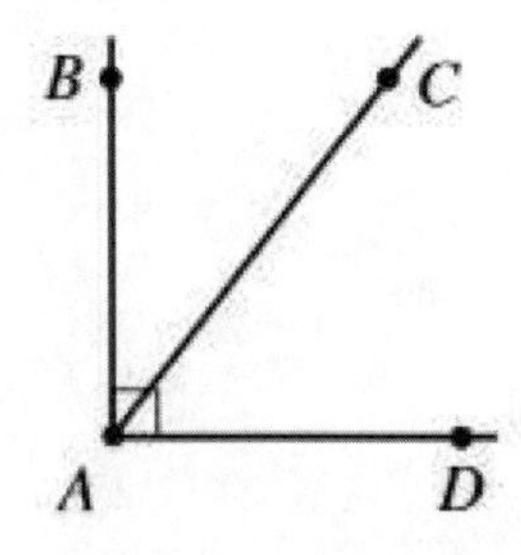

a. $(x - 125)°$
b. $(125 - x)°$
c. $(125 + x)°$
d. $(x - 55)°$
e. $(55 - x)°$

42. In the figure below, where $\Delta ABC \sim \Delta XYZ$, lengths given are in centimeters. What is the perimeter, in centimeters, of ΔXYZ ?

PE 2 42

(Note: The symbol ~ means "is similar to.")

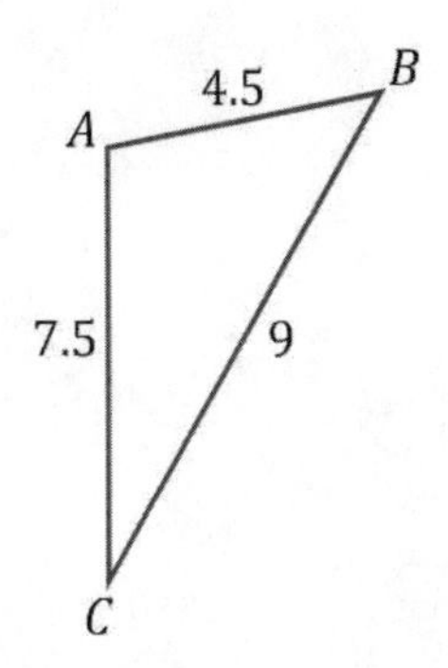

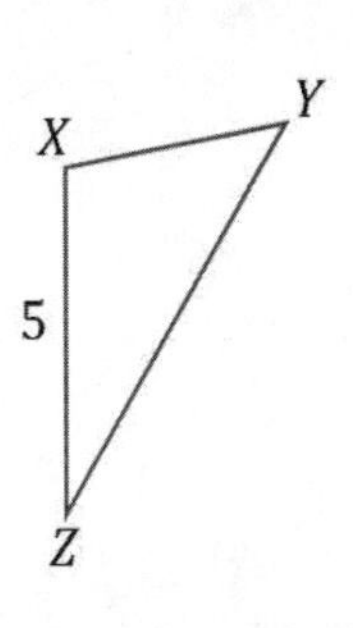

a. 12

b. $13\frac{1}{2}$

c. 14

d. 16

e. $21\frac{3}{4}$

43. The figure below is composed of square $LMOP$ and equilateral triangle ΔMNO. The length of $\overline{LP}$ is 8 inches. What is the perimeter of $LMNOP$ in inches?

PE 2 43

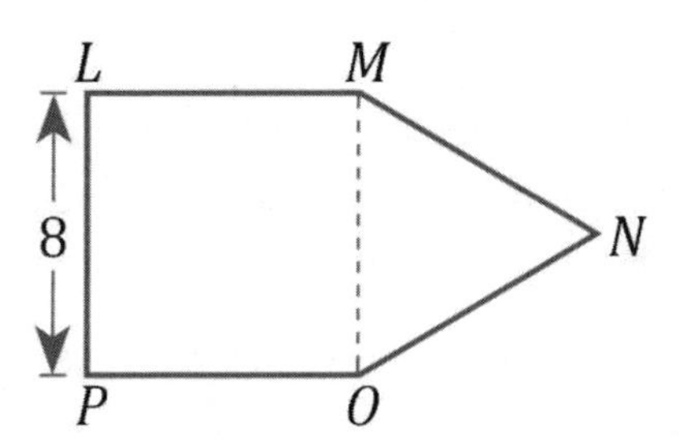

a. 24
b. 32
c. 40
d. 56
e. 60

44. Curt plans to paint the 9-foot-high rectangular walls of his bedroom and before he buys paint he needs to know the area of the wall surface to be painted. Two walls are 12 feet wide, and the other 2 walls are 9 feet wide. The combined area of the 1 window and the 1 door in his room are 65 square feet. What is the area, in square feet, of the wall surface Curt plans to paint?

PE 2 44

a. 189
b. 313
c. 324
d. 366
e. 378

45. The length of a rectangle is one quarter of the length of a larger rectangle. The 2 rectangles have the same width. The area of the smaller rectangle is A square units. The area of the larger rectangle is dA square units. Which of the following is the value of d?

PE 2 45

a. 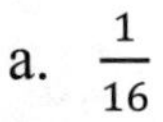 $\frac{1}{16}$

b. $\frac{1}{4}$

c. 1

d. 4

e. 16

46. Dano wants to find the volume of a solid figurine. He fills a rectangular container 10 cm long, 7 cm wide, and 13 cm high with water to a depth of 6 cm. Dano totally submerges the figurine in the water. The height of the water with the submerged figurine is 8.2 cm. Which of the following is closest to the volume, in cubic centimeters, of the figurine?

PE 2 46

a. 154
b. 200
c. 286
d. 420
e. 574

Use the following information to answer the next three questions.

Mika is developing a proposal for a new playground. The figure below shows her scale drawing of the proposed playground with 3 side lengths and the radius of the merry-go-round given in inches. In Mika's scale drawing, 1 inch represents 2.5 feet.

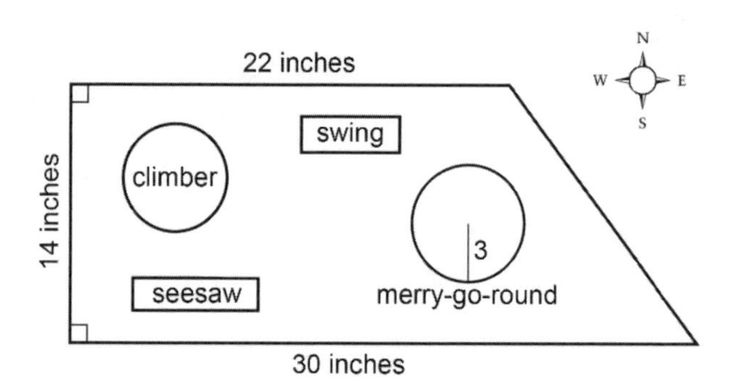

47. What is the area, in square inches, of the scale drawing of the playground?

PE 2 47

a. 308
b. 364
c. 420
d. 476
e. 728

48. Mika's proposal includes installing a fence on the perimeter of the playground. What is the perimeter, to the nearest foot, of the playground?

PE 2 48

a. 66
b. 82
c. 165
d. 205
e. 220

49. The length of the south side of the playground is what percent of the length of the north side?

PE 2 49

a. 108%
b. 116%
c. $136\frac{4}{11}\%$
d. $173\frac{1}{3}\%$
e. 225%

50. In ΔPQR, the length of $\overline{PQ}$ is 6 inches, and the length of $\overline{QR}$ is $\sqrt{34}$ inches. If it can be determined, what is the length, in inches, of $\overline{PR}$?

PE 2 50

a. 6
b. $\sqrt{34}$
c. $\sqrt{40}$
d. $\sqrt{70}$
e. Cannot be determined from the given information

51. For right triangle ΔABC below, what is $\cos \angle A$?

PE 2 51

a. $\frac{11}{14}$

b. $\frac{14}{11}$

c. $\frac{\sqrt{75}}{11}$

d. $\frac{11}{\sqrt{75}}$

e. $\frac{\sqrt{75}}{14}$

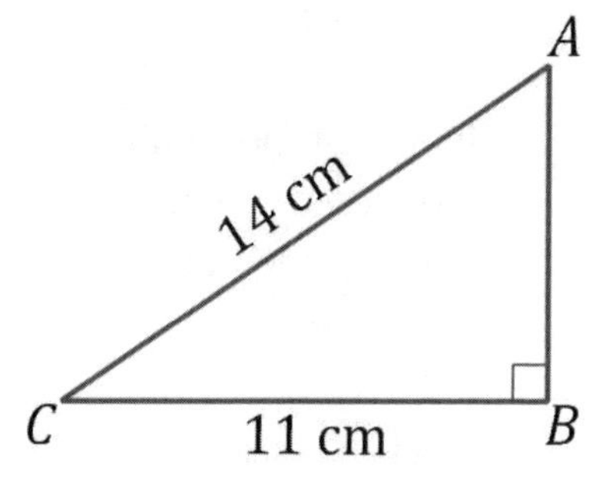

52. The graph of the trigonometric function $y = \frac{1}{2}\cos(2x)$ is shown below.

PE 2 52

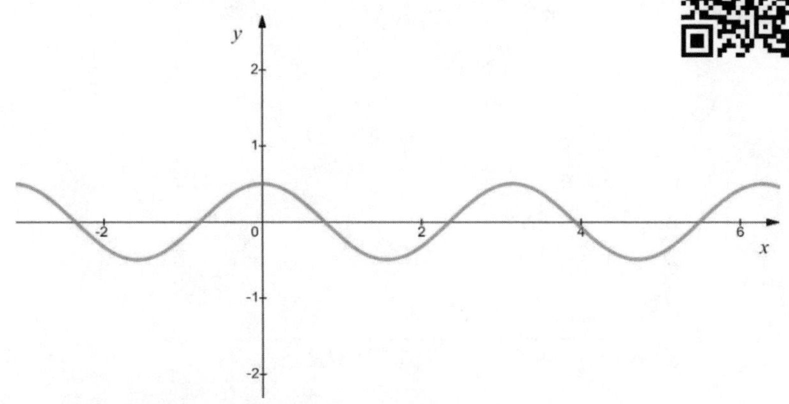

The function is:

a. even (that is, $f(x) = f(-x)$ for all x).
b. odd (that is, $f(-x) = -f(x)$ for all x).
c. neither even nor odd.
d. the inverse of a cotangent function.
e. undefined at $x = \pi$

53. Triangle ΔABC is shown in the figure below. The measure of $\angle B$ is 48°, $AB = 19$ cm, and $BC = 15$ cm. Which of the following is the length, in centimeters, of $\overline{AC}$?

PE 2 53

(Note: For a triangle with sides of length a, b and c and opposite angles $\angle A, \angle B$, and $\angle C$, respectively, the law of sines states $\frac{\sin \angle A}{a} = \frac{\sin \angle B}{b} = \frac{\sin \angle C}{c}$ and the law of cosines states $c^2 = a^2 + b^2 - 2ab\cos\angle C$.)

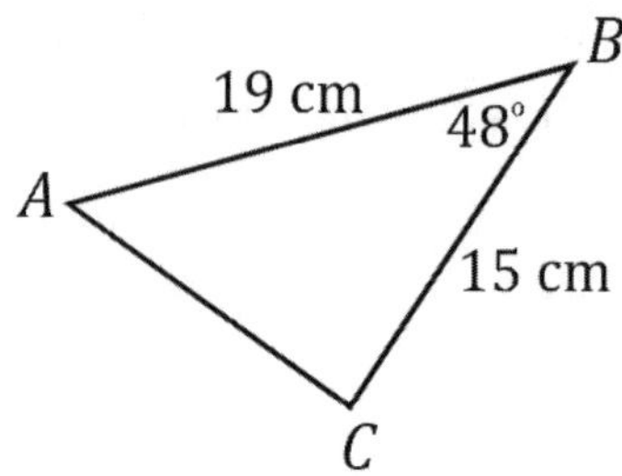

a. $15 \sin 48°$
b. $19 \sin 48°$
c. $\sqrt{19^2 - 15^2}$
d. $\sqrt{15^2 + 19^2}$
e. $\sqrt{15^2 + 19^2 - 2(15)(19) \cos 48°}$

54. A function $f(x)$ is defined as $f(x) = -4x^3$. What is $f(-2)$?

PE 2 54

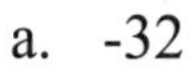

a. -32
b. -24
c. 24
d. 32
e. 512

55. Given $f(x) = 6x - 7$ and $g(x) = x^2 - 2x + 1$, which of the following is an expression for $f(g(x))$?

PE 2 55

a. $6x^2 - 12x - 6$
b. $6x^2 - 12x - 1$
c. $6x^2 - 2x - 6$
d. $36x^2 - 12x - 64$
e. $36x^2 - 96x - 64$

56. For a population that grows at a constant rate of $r\%$ per year, the formula

PE 2 56

$$P(t) = A\left(1 + \frac{r}{100}\right)^t$$

models the population t years after an initial population of A people is counted.

The population of the city of Calgary was 568,000 in 1980. Assume the population grows at a constant rate of 4.8% per year. According to this formula, which of the following is an expression for the population of Calgary in the year 2000?

a. $568{,}000(5.8)^{20}$
b. $568{,}000(1.48)^{20}$
c. $568{,}000(1.048)^{20}$
d. $(568{,}000 \times 1.48)^{20}$
e. $(568{,}000 \times 1.048)^{20}$

57. The table below shows the price of different quantities of avocados at Planet Organica. What is the least amount of money needed to purchase exactly 22 avocados if the bags must be sold intact and there is no tax charged for avocados?

PE 2 57

Number of avocados	1	Bag of 6	Bag of 12
Total price:	\$3.25	\$13.00	\$22.75

a. \$39.00
b. \$45.50
c. \$48.75
d. \$52.00
e. \$71.50

58. A construction company builds 3 different models of houses (A, B, and C). They order all the bath tubs, shower stalls, and sinks for the houses from a certain manufacturer. Each model of house contains different numbers of these bathroom fixtures. The tables below give the number of each kind of these fixtures required for each model and the cost to the company, in dollars, of each type of fixture.

PE 2 58

Fixture	Model		
	A	B	C
Bathtubs	1	1	2
Shower stalls	2	2	2
Sinks	1	3	4

Fixture	Cost
Bathtub	$450
Shower stall	$275
Sink	$130

The company plans to build 5 A's, 4 B's and 2 C's. What will the cost to the company be to put the required number of bathroom fixtures in all of these houses?

a. $4,490
b. $8,980
c. $12,500
d. $15,150
e. $49,390

59. As part of a lesson on motion, students observed a ball rolling at a constant rate along a straight track. As shown in the chart below, they recorded the distance, d inches, of the ball from a reference point at 1-second intervals from 0 to 5 seconds.

PE 2 59

t	0	1	2	3	4	5
d	12	16	20	24	28	32

Which of the following equations represents this data?

a. $d = t + 12$
b. $d = 4t + 8$
c. $d = 4t + 12$
d. $d = 12t + 4$
e. $d = 16t$

60. The fluctuation of water depth at a harbor is shown in the graph below. One of the following values gives the positive difference, in feet, between the greatest water depth and the least water depth shown in this graph. Which value is it?

PE 2 60

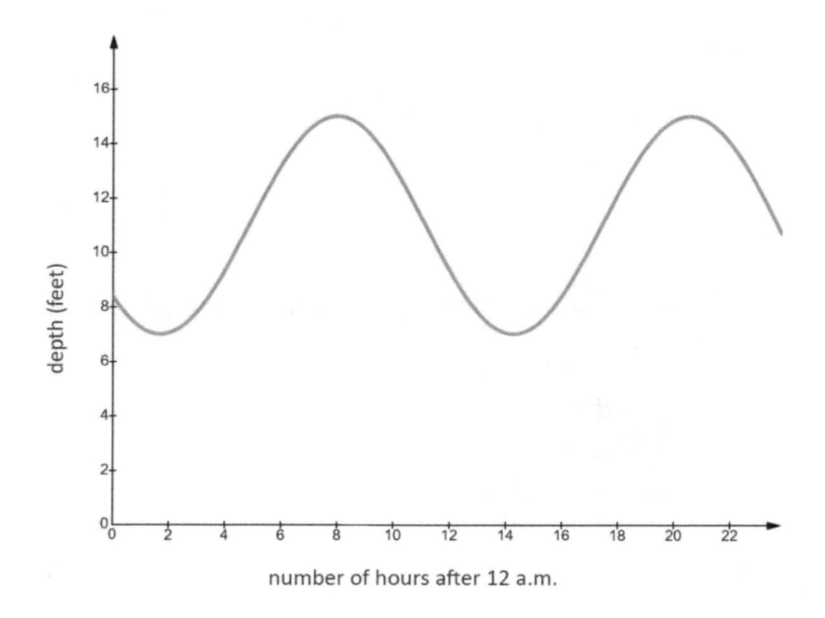

a. 7
b. 7.5
c. 8
d. 12
e. 15

Solutions

ACT Exam Review Manual Answer Key

Pre-Algebra

1. C
2. C
3. C
4. C
5. E
6. C
7. B
8. D
9. A
10. D
11. E
12. D
13. D
14. E
15. D
16. E
17. A
18. A
19. D
20. D
21. C
22. E
23. E
24. C
25. D

Algebra 1

1. D
2. C
3. B
4. E
5. D
6. A
7. B
8. B
9. C
10. D
11. D
12. D
13. A
14. D
15. C
16. A
17. C
18. E

Algebra 2

1. B
2. D
3. C
4. B
5. B
6. C
7. B
8. D
9. C
10. C
11. C
12. D
13. A
14. C

Algebra 3

1. D
2. A
3. A
4. B
5. A
6. A
7. C
8. D
9. E
10. C
11. E
12. A
13. E
14. A

Algebra 4

1. E
2. E
3. C
4. D
5. D
6. C
7. C

Geometry 1

1. D
2. A
3. A
4. C
5. C
6. D
7. D
8. B
9. B
10. D
11. C
12. B
13. A
14. C
15. C
16. E
17. B
18. B
19. D
20. E
21. A
22. E
23. C

Geometry 2

1. B
2. C
3. D
4. B
5. B
6. B
7. D
8. C
9. E
10. C
11. B

Geometry 3

1. D
2. D
3. A
4. E
5. B
6. C
7. B
8. B
9. E
10. E
11. C
12. C
13. E
14. B
15. B
16. E
17. A
18. C
19. A
20. D
21. B
22. C
23. B

Geometry 4

1. E
2. A
3. D
4. A
5. E
6. E
7. D
8. D
9. C
10. A
11. B
12. D
13. E
14. B
15. B

Functions 1

1. A
2. B
3. D
4. A
5. B
6. D

Functions 2

1. C
2. E
3. C
4. B
5. D
6. C
7. C

Probability

1. A
2. C
3. E
4. C
5. C
6. D
7. C

Statistics

1. B
2. C
3. D
4. D
5. D
6. D
7. E
8. C
9. C
10. D

Practice Exam 1

1. C
2. E
3. D
4. E
5. A
6. B
7. A
8. D
9. B
10. B
11. D
12. D
13. D
14. E
15. C
16. B
17. D
18. A
19. B
20. C
21. C
22. C
23. B
24. D
25. B
26. C
27. E
28. D
29. A
30. C
31. C
32. E
33. D
34. B
35. B
36. C
37. C
38. E
39. D
40. C
41. D
42. B
43. C
44. D
45. D
46. E
47. B
48. E
49. D
50. E
51. C
52. E
53. B
54. C
55. B
56. B
57. B
58. A
59. E
60. D

Practice Exam 2

1. E
2. C
3. B
4. E
5. C
6. C
7. B
8. D
9. E
10. B
11. C
12. E
13. C
14. B
15. E
16. A
17. B
18. D
19. E
20. C
21. C
22. B
23. C
24. D
25. A
26. E
27. C
28. A
29. D
30. E
31. E
32. B
33. D
34. B
35. A
36. A
37. B
38. C
39. C
40. B
41. B
42. C
43. C
44. B
45. D
46. A
47. B
48. D
49. C
50. E
51. E
52. A
53. E
54. D
55. B
56. C
57. C
58. D
59. C
60. C

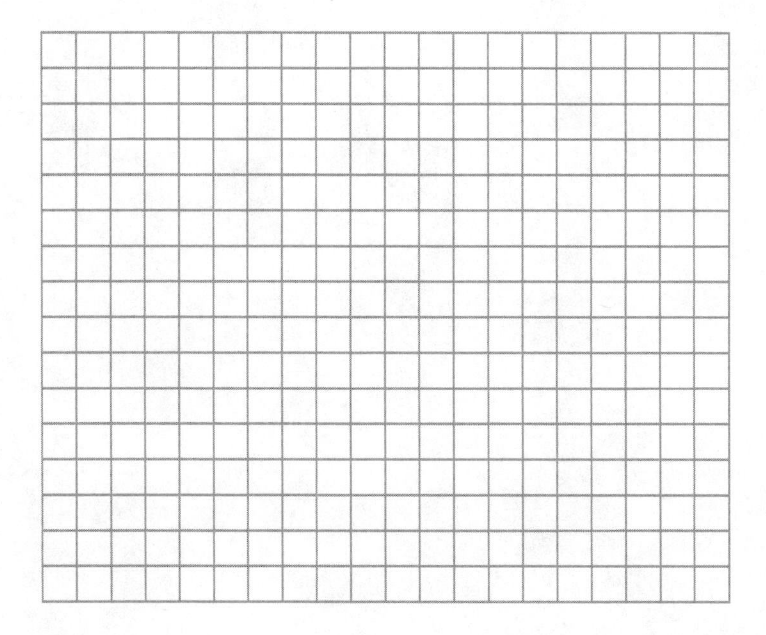

NOTES:

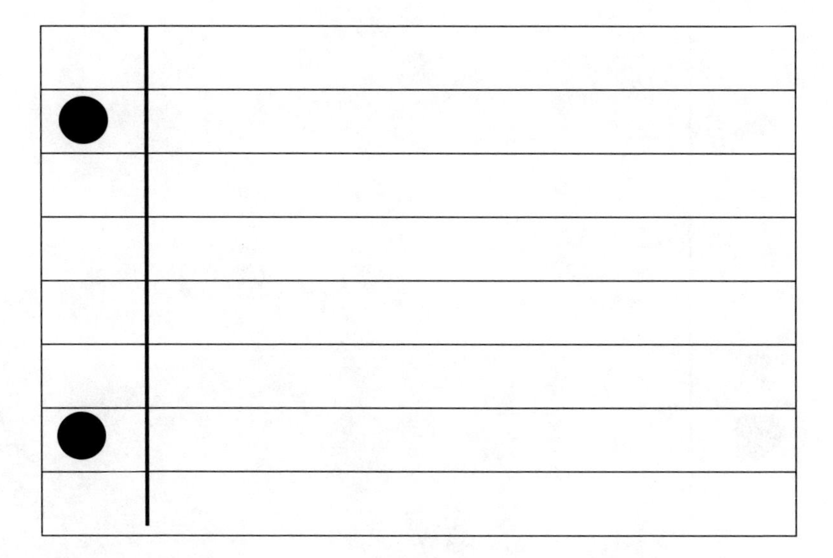

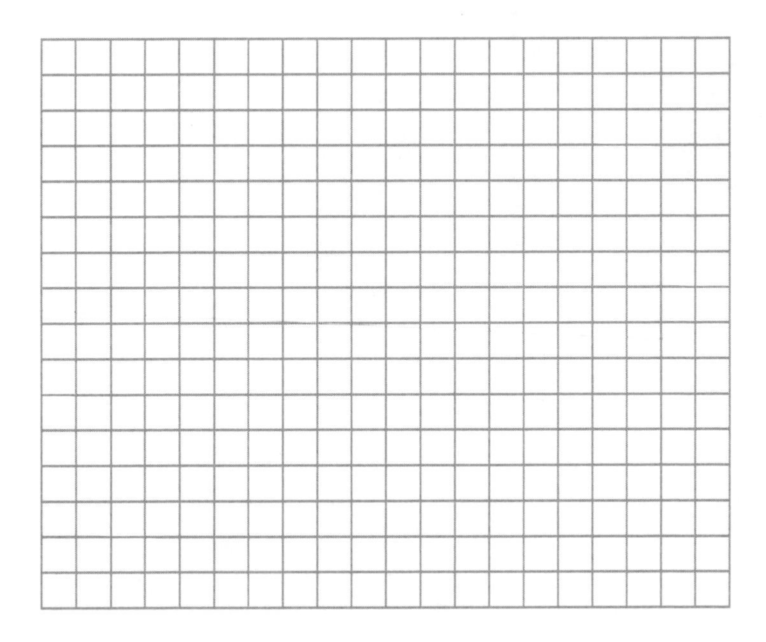

NOTES:

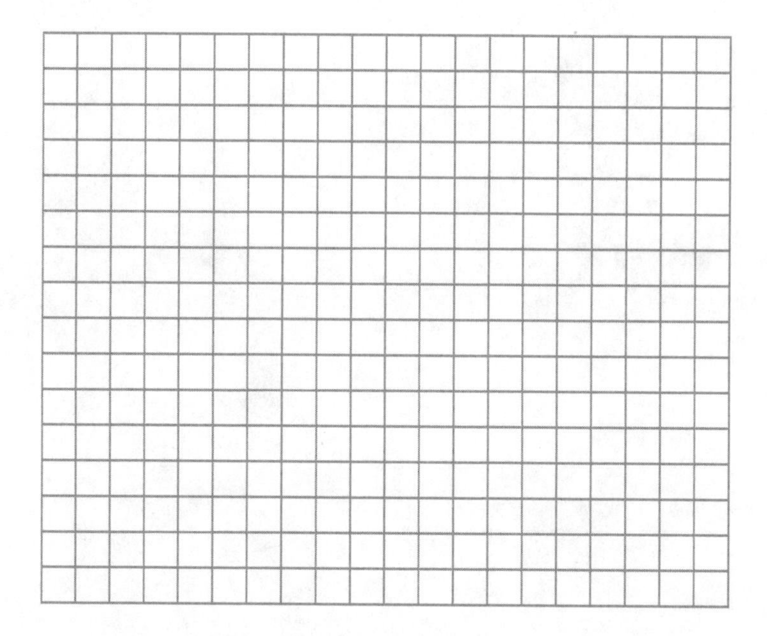

NOTES:

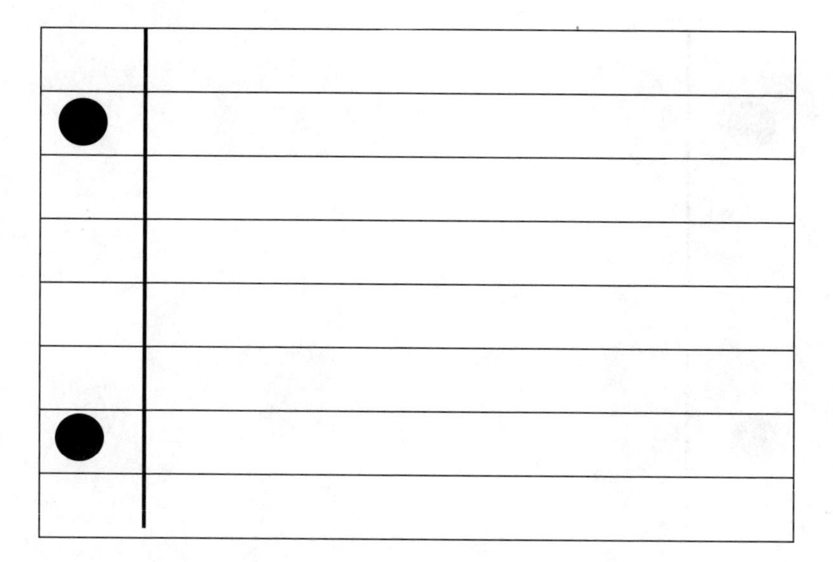

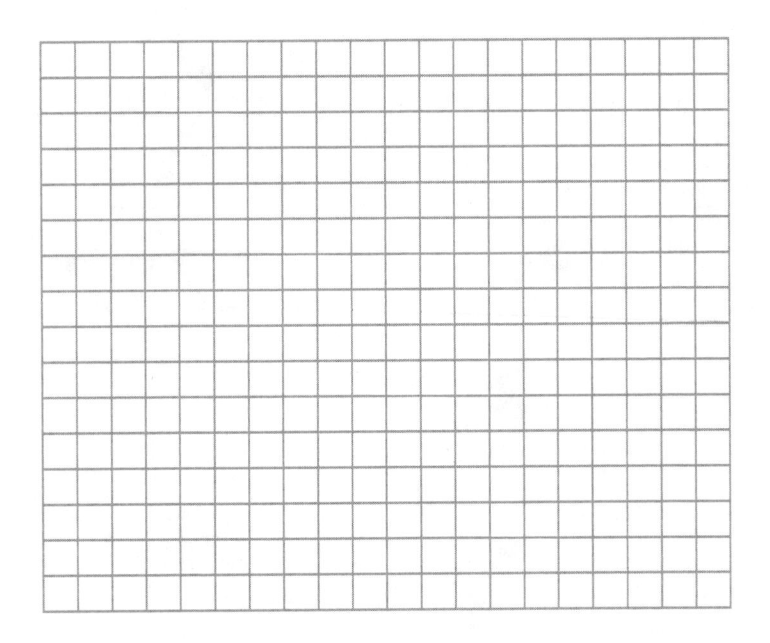

NOTES:

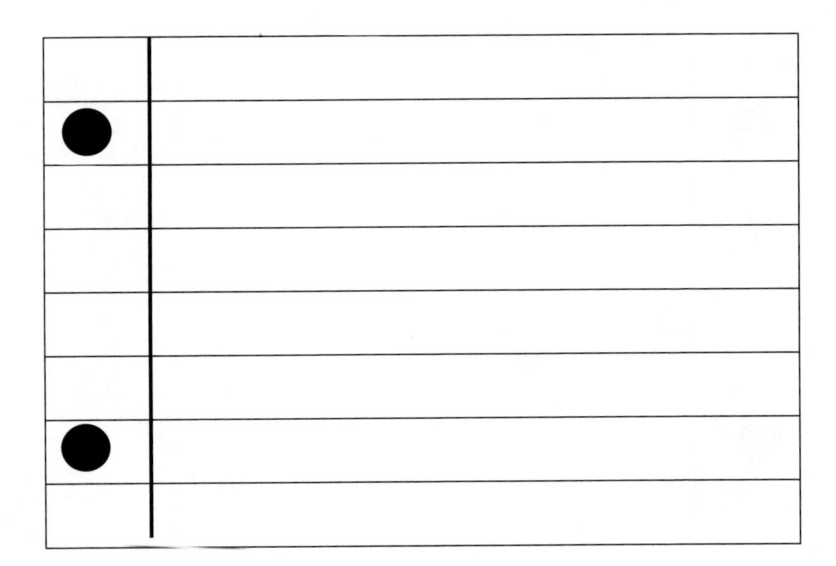

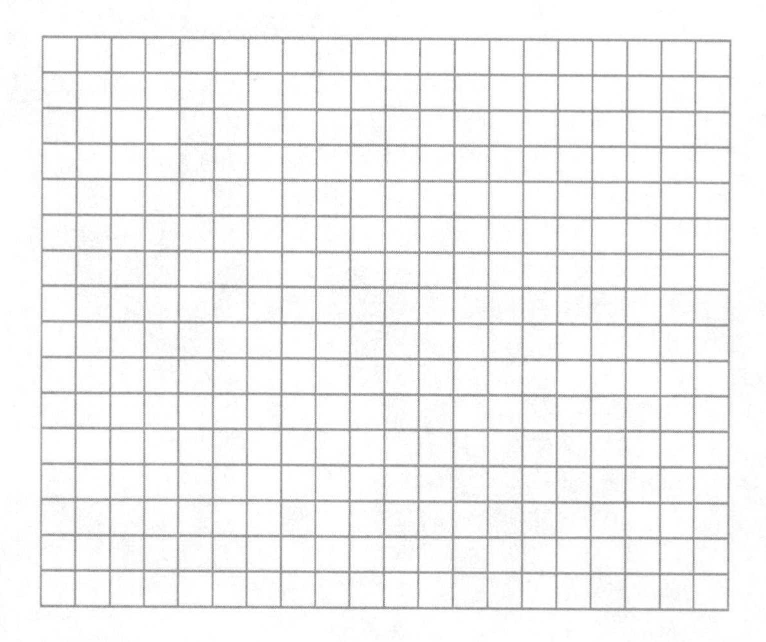

NOTES:

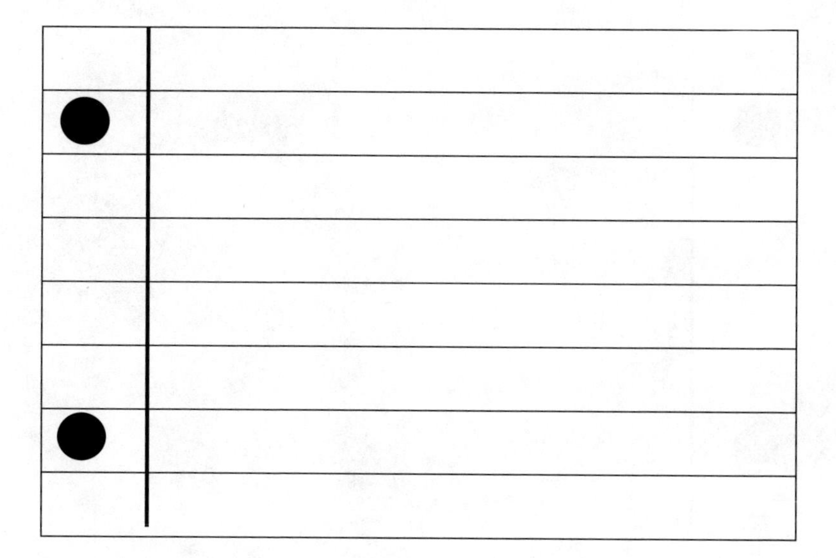

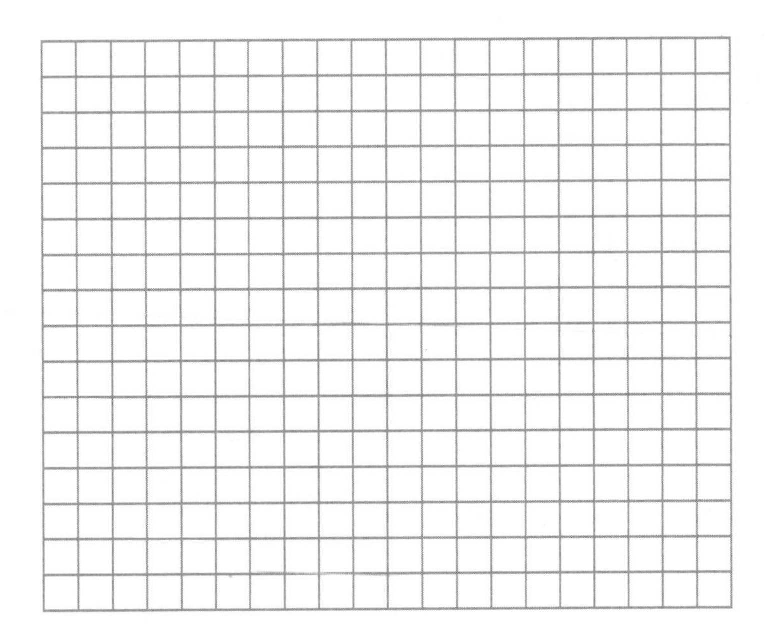

NOTES:

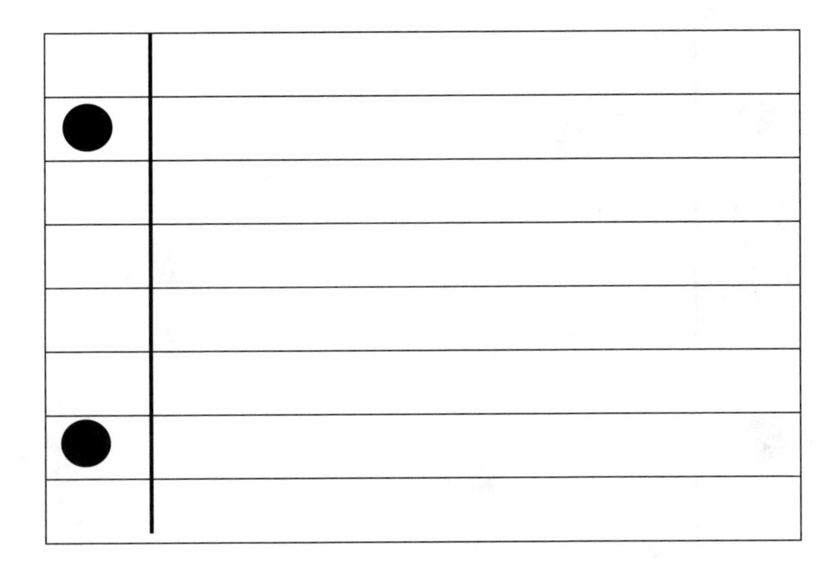

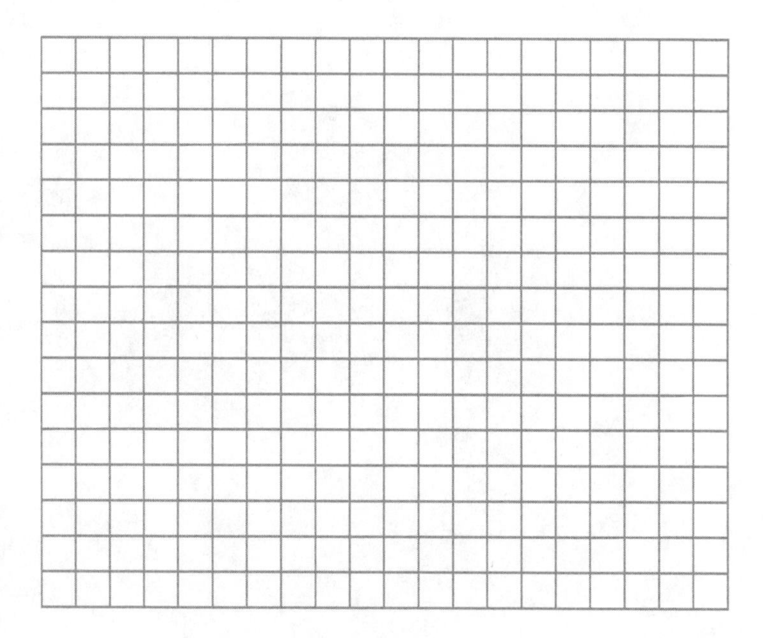

NOTES:

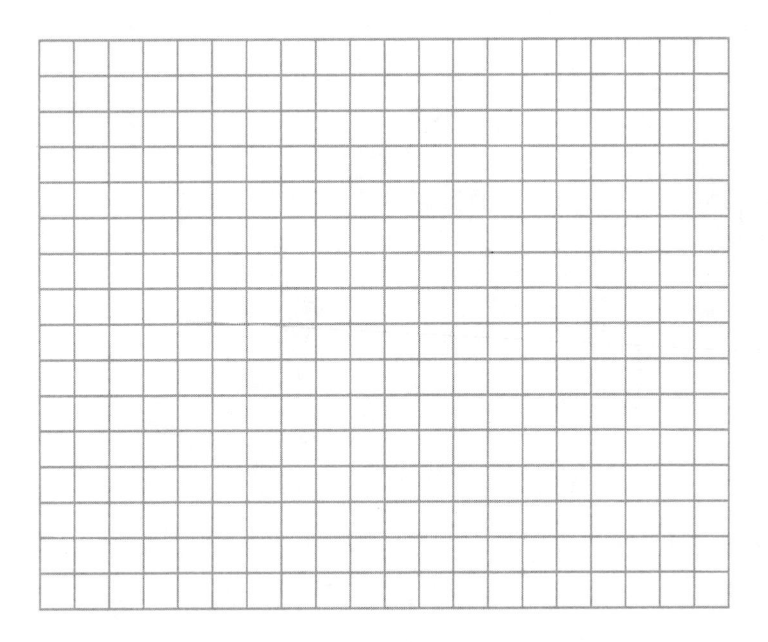

NOTES:

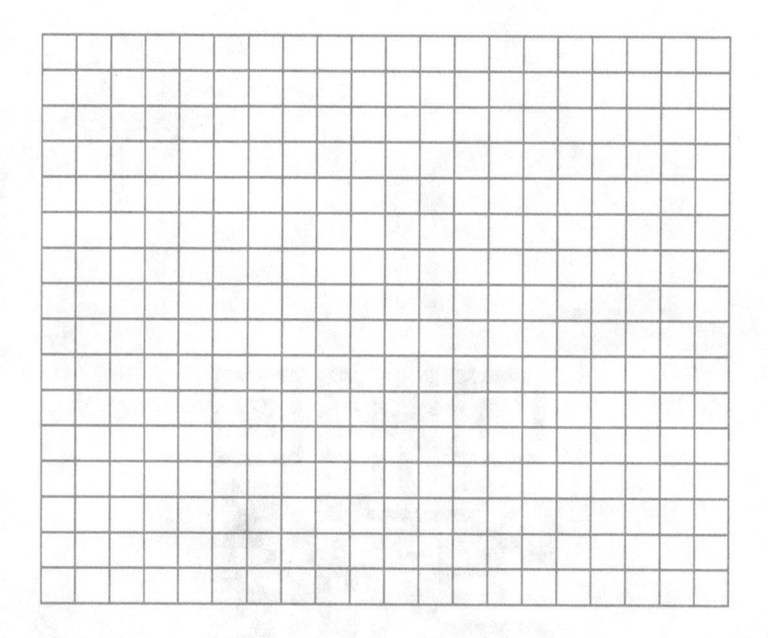

NOTES:

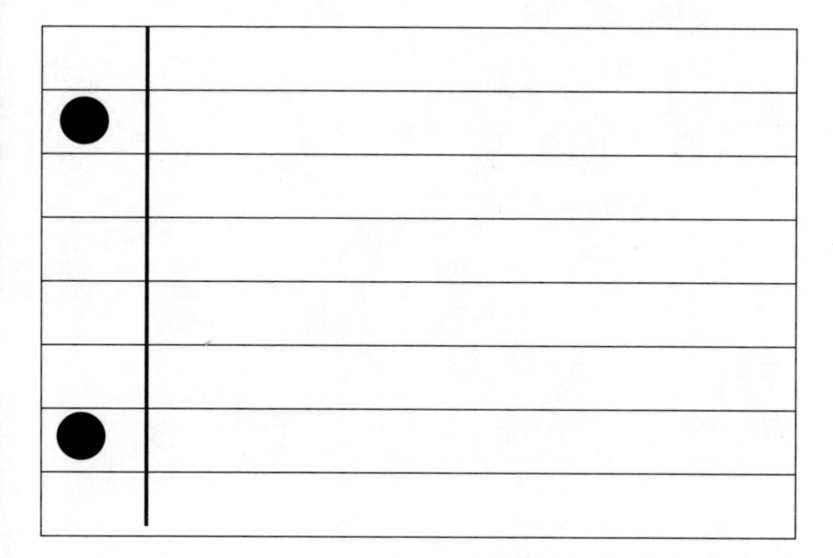